D1126581

BASIC TRACK & FIELD BIOMECHANICS

Tom Ecker

First published in 1985 by Tafnews Press
Book Division of Track & Field News, Inc.
P.O. Box 296, Los Altos, California 94022 USA

ISBN 0-911521-16-X

Printed in the United States of America

My sincere thanks to Elmer Blasco and Jack Griffith of *Athletic Journal* for providing the majority of the photographs, to Sharon Begley for her artwork, to Robin Barnes for her tireless help in preparing the manuscript, and to Karen and Mandie for their occasional tolerance.

–Tom Ecker

Cover photograph: Jeff Johnson

CONTENTS

Preface . . . 7
PART I–Basic Concepts of Biomechanics . . . 9
1. Linear Motion . . . 11
   Speed and velocity . . . 12
   Component velocities. . . . 12
   Acceleration and deceleration . . . 16
   Acceleration due to gravity . . . 16
2. Air Resistance . . . 18
   Running . . . 18
   Jumping . . . 19
   Throwing. . . . 19
3. Center of Mass. . . . 21
   Stability and balance . . . 23
   Balance and motion. . . . 24
4. Curves of Flight . . . 29
   Parabolic curves . . . 29
   Aerodynamic curves . . . 32
   Optimum angles for achieving horizontal distance . . . 32
   Throwing implements . . . 34
   Horizontal jumpers . . . 35
5. Inertia and Momentum . . . 36
6. Ground Reaction . . . 38
   Impulse . . . 38
7. Rotary Motion . . . 41
   Axes . . . 41
   Primary axes. . . . 41
   Nutation . . . 43
   Secondary axes . . . 43
8. Rotary Inertia and Rotary Momentum . . . 46
   Rotary momentum . . . 48
   Conservation of rotary momentum . . . 48
9. Rotation in the Air . . . 52
   Rotation originating on the ground. . . . 52

Rotation originating in the air . . . 55
10. Centripetal and Centrifugal Forces . . . 62

PART II–The Running Events . . . 67
11. Basic Biomechanics of Running . . . 69
The running stride . . . 69
Stride length and stride frequency . . . 72
Forward lean in running . . . 74
Hurdling . . . 76
12. Sprinting . . . 78
Improving sprinting speed . . . 78
Sprint starting . . . 79
Force against the blocks . . . 79
Body mass . . . 81
Acceleration in sprinting . . . 81
Air resistance . . . 84
Finishing dip . . . 84
Selecting starting blocks . . . 85
Coaching pointers . . . 87
13 Endurance Running . . . 89
Coaching Pointers . . . 90
14. Hurdling . . . 91
High hurdling . . . 91
Low and intermediate hurdle clearance . . . 97
Hurdle selection . . . 99
Coaching pointers . . . 101
15. Relay Racing . . . 102
The 4x100m relay . . . 102
Determining sprint go-mark distances . . . 105
The 4x400m relay . . . 108
Coaching pointers . . . 108

PART III–The Jumping Events . . . 109
16. Basic Biomechanics of Jumping . . . 111
Ground reaction . . . 111
Takeoff angles . . . 112
Flight curves . . . 112
17. High Jumping . . . 113
The approach . . . 113
The takeoff . . . 114
Bar clearance . . . 117
Coaching pointers . . . 118
18. Pole Vaulting . . . 119
The approach run and pole plant . . . 119
The takeoff and swing-up . . . 121
The pullup, turn and pushoff . . . 124
Bar clearance . . . 124

Vaulting pole selection . . . 125
Pole vault pit selection . . . 130
Coaching pointers. . . . 134
19. Long Jumping. . . . 135
The takeoff . . . 135
In the air . . . 139
The landing . . . 145
Coaching pointers. . . . 146
20. Triple Jumping . . . 147
The hop . . . 147
The step . . . 149
The jump. . . . 151
Coaching pointers. . . . 152

PART IV–The Throwing Events . . . 155
21. Basic biomechanics of throwing. . . . 157
Speed of release . . . 158
Ground reaction. . . . 158
Angle of release . . . 159
Flight curves . . . 159
22. Shot Putting . . . 161
Strength and size . . . 161
Angle of release . . . 164
Conventional technique . . . 165
Rotational technique . . . 169
Shot selection . . . 169
Coaching pointers. . . . 171
23. Discus Throwing . . . 172
Speed of release . . . 172
Ground reaction. . . . 176
Angle of release . . . 178
Preliminary swings . . . 178
The turn . . . 178
The delivery . . . 180
Flight of the discus . . . 181
Discus selection . . . 183
Coaching Pointers . . . 183
24. Hammer Throwing . . . 185
Centripetal and centrifugal forces. . . . 185
Angle and speed of release . . . 187
Preliminary swings . . . 187
The turns. . . . 189
The delivery . . . 190
Hammer selection. . . . 190
Coaching pointers. . . . 193
25. Javelin Throwing . . . 194
Speed of release . . . 195

Angle of release 196
Gripping the javelin 198
The runup 198
The delivery 199
After the release 201
Javelin selection 202
Coaching pointers 203

Bibliography 204

PHOTO CREDITS

Fig. 12 . . . . Rhein-Ruhr-Foto: Gustav Schroeder
Fig. 13 . . . . Jeff Johnson
Fig. 35 . . . . Jeff Johnson
Fig. 49 . . . . Jeff Johnson
Fig. 52 . . . . Victor Sailer
Fig. 53 . . . . Erik Hill
Fig. 54 . . . . Diane Johnson
Fig. 55 . . . . Walley W. Brown
Fig. 57 . . . . Maurice Wilson
Fig. 60 . . . . Bill Leung, Jr.
Fig. 70 . . . . Don Chadez
Fig. 72 . . . . Don Chadez
Fig. 73 . . . . Jeff Johnson

PREFACE

Twenty years ago, in Orlando, Florida, I presented my first track & field biomechanics lecture. The audience, made up of Florida coaches, was small but enthusiastic.

The science wasn't called biomechanics then. It was mechanics, a name that had come to us from England. Since mechanics reminded me of car repairs, I decided to rename the science dynamics, thinking the name might stick. But soon biomechanics took over as the official name.

The projector for the first lecture was a used Specto Analyzer, manufactured in Great Britain. Primitive by today's standards, it was the only analyzing projector available. The turntable I used for demonstrating action-reation and conservation of rotary momentum was a stripped-down lazy susan.

The main topic of that first talk was straddle high jumping, a technique which I felt was grossly misunderstood. Now it is not just misunderstood, it is virtually extinct.

I had been privileged to be able to listen to Geoffrey Dyson at coaching clinics during the previous two summers and had the extreme good fortune to be able to spend a great deal of time discussing this "new" science with the world's greatest collector and disseminator of track & field knowledge, Fred Wilt. Dyson and Wilt, and later, Jim Hay, changed my ideas about almost everything. They made me realize how much there is to learn after you know it all.

Over the years, my biomechanics lectures have evolved into a rapid-fire, two-hour "dog and pony show" (sprinkled with visual surprises and occasional craziness), that has taken me before coaching groups in 34 states and 9 foreign countries. The props for "the show" include a remote-controlled 16mm analyzer projector, a sophisticated hand-made portable turntable, and dozens of "gadgets" designed to demonstrate biomechanical concepts.

The goal of this book is the same as the goal of that first lecture 20 years ago—and every lecture since: to translate from the language of the scientist to the language of the coach, athlete and spectator. Some of the language of the scientist has been altered for better understanding; some concepts have been simplified.

When my first track & field book was published 25 years ago, the late George Gibson of Louisville, Kentucky, wrote in the foreword of the obligation we all have to "repay the debts we owe to track & field athletics." For me, that debt will always be there. This book is partial payment.

—Tom Ecker

P.S. The English language, although renowned for its versatility in most repects, has one great failing. It does not provide third-person singular pronouns which cover both sexes. The plurals *they, them,* and *their* do not provide a problem, since they refer to both sexes, but that is not the case with the singular pronouns. After many days of struggling with various alternatives, all of which produced awkward phrasing, it was decided—out of necessity—to use the generic pronouns *he, him,* and *his* throughout this book. In most cases, however, the pronouns *she, her,* and *hers* can be substituted.

PART I

BASIC CONCEPTS OF BIOMECHANICS

1 LINEAR MOTION

There are two types of motion—linear and rotary.

Linear motion, also called *translation,* is motion along a straight or curved line, such as the path of an automobile traveling along a highway. All parts of the automobile's frame and body move the same distance, in the same direction, in the same time.

Rotary motion, which will be covered in detail in Chapters 7-10, is turning motion in a circle (or arc) around an axis, such as the motion of one of the automobile's wheels around its axle.

Every track & field event has both linear motion and rotary motion. The linear motion of a runner is coupled with the rotary motions of swinging arms and driving legs. Jumpers attempt to achieve great height or distance (depending upon the event) with linear motion and are aided by various rotary movements. Throwers use rotational movements to accelerate and release their implements with linear motion.

In track & field, there are few, if any, examples of *pure* linear motion, since linear movements require that all parts of an object (or person) move the exact same distance in the same time. A sprinter's trunk, therefore, does not have pure linear movement, since it bends backward and forward slightly as the sprinter races down the track. However, for our purposes here, we can say that the sprinter's trunk has *apparent* linear motion.

Speed And Velocity

The terms *speed* and *velocity* are often used interchangeably, but they are not the same. Speed is the rate of motion of a body; velocity is the body's speed in a particular direction. A sprinter's speed may be 30 feet per second, but if that speed is known to be in a particular direction (whether the direction is horizontal or not), then the 30fps must be stated as velocity.

The speed (rate of motion) of a body is measured in units of length and time, such as feet per second (fps), meters per second (mps), miles per hour (mph), or kilometers per hour (kph).

Since the velocity of a body has both magnitude and direction, these factors can be illustrated diagramatically by drawing a straight arrow in the direction of the velocity. The length of the arrow represents the magnitude of the velocity, drawn to some selected scale.

When there are two velocities at the same time, one forward and one upward, as is often the case in track & field motions, the combination of the two *component velocities* produces a *resultant velocity.* The lengths of the horizontal and vertical arrows represent the comparison of the magnitudes of the component velocities. (See Figure 1.)

Component Velocities

When an athlete leaves the ground and is free in the air (as in the jumping events, the hurdles, or even during an individual running stride), or when throwing implements are released in the air, two separate velocities have been imparted to the athlete or the implement–one horizontal and one vertical. Separately, these component velocities would provide either straight-forward motion or straight-up lift, but together they produce a resultant velocity. It is this resultant

Figure 1. The horizontal arrow represents the magnitude of the long jumper's horizontal velocity. The vertical arrow represents the magnitude of the vertical velocity.

velocity that determines the angle and the speed of the athlete's takeoff or of the release of the implement. (See Figure 2.)

The component velocities should be thought of as separate, even though it is not always easy to visualize them separately. The sprint start, for example, gives little appearance of either a horizontal or a vertical velocity as the sprinter leaves the blocks.

An event in which the separate velocities are more obvious is the conventional shot put. The putter begins with velocity that is horizontal, and then adds the vertical velocity separately. The *shift* across the circle provides most of the horizontal component. The *lift* of the putter provides most of the vertical component. (See Figure 3.) The resultant

Figure 2. The horizontal and vertical velocities established earlier determine the angle and the speed of the jumper's takeoff and the putter's release.

Figure 3. The conventional shot putter moves the shot horizontally first, and then adds a powerful vertical velocity before release.

velocity, a combination of shift and lift, determines the release velocity and the angle of release, no matter how much the putter might want to raise or lower the angle at the time of release. (A fast shift and slow lift produces a low release angle; a slow shift and fast lift produces a high release angle.)

Acceleration And Deceleration

In sports, velocities are seldom constant, and so we must think in terms of gaining velocity and losing velocity. Comparing the velocity of a body at one moment with its velocity a moment before or a moment later determines whether the body is accelerating or decelerating.

Acceleration is rate of change of velocity. An increase in velocity is called positive acceleration. A decrease is called negative acceleration, or deceleration.

In sprint starting—because the sprinter is accelerating from zero velocity—the positive change in velocity with the first stride is especially great and thus produces the greatest acceleration. There is still acceleration with the second stride from the blocks since the sprinter continues to move faster, but the increase in velocity is not as great as with the first stride. In other words, while velocity is increasing with each stride, the acceleration is decreasing.

The sprinter continues to accelerate until the legs can no longer move faster than the ground is moving beneath the feet (usually the limit is about six seconds). Then a steady velocity is maintained briefly (with zero acceleration) before a gradual deceleration begins.

Acceleration Due to Gravity

Until the time of Galileo it was believed that heavier objects fell faster than light objects. We now know (disregarding air resistance for the moment) that all objects, regardless of weight, fall with a constant acceleration of approximately 32 feet per second for every second they fall.

If you were to drop a marble and a cannonball from the

same height at the same time, the two would accelerate equally (at a constant rate of 32 fps^2), and they would hit the ground at the same time. If the marble and the cannonball were welded together and then dropped, side by side, the cannonball would not rotate downward into a lower position. The two objects would still drop at the same rate, with no change of relative positions (again, ignoring whatever effects air resistance might have on different-sized bodies).

Similarly, if you were to hold a carpenter's hammer in a horizontal position and let it fall to the ground (from any height), even though the head of the hammer is considerably heavier than the handle, the hammer would not rotate into a head-down position.

An arrow shot into the air (or a bomb dropped from an airplane) cannot be made to land point first by making the point heavier, since the heavy point falls no faster than the lighter tail. There must be fins to catch the air, forcing the tail of the arrow (or bomb) to fall behind.

A javelin does not land point first because its point is heavier than its tail; it does so because the tail is designed to catch more air than the point, causing the javelin to rotate into a point-down position.

The same is true of the human body free in the air–be it pole vaulter or long jumper.. Unless the body is set in rotation before leaving the ground, all parts of the body fall at the same rate of speed.

2
AIR RESISTANCE

When analyzing track & field techniques, air resistance is usually ignored or disregarded in cases in which the resistance is so slight that it has no appreciable effect on performance. Examples include the flight paths of shots and hammers, and even the flight paths of jumpers and hurdlers when there is no wind to complicate matters.

There are times, however, when the effects of air resistance (and wind resistance) must be considered when analyzing technique.

Running

In the running events, air resistance effects posture slightly. As a runner's velocity increases, there is an increase in the air resistance he must face. To counteract the tendency to fall backwards because of this resistance, forward body lean must increase. But it is a small degree of lean when compared with the effects of acceleration and deceleration on the body's posture.

When a sprinter faces a headwind, the increased air resistance requires a more acute forward lean in order to continue running in balance. And, of course, the headwind reduces the runner's horizontal velocity.

Conversely, a following wind allows the sprinter to face reduced air resistance, which increases his horizontal velocity.

However, there is always some air resistance, even when there is a strong following wind (unless, of course, the velocity of the following wind is greater than that of the sprinters, which is very unlikely).

In races of one lap or more in which runners cover equal distances facing a headwind and enjoying a tailwind, the headwind has a greater negative effect on the runner's energy and speed than the tailwind's positive effect. The result is that more time is lost facing the wind than is gained running with it, resulting in a net loss in the time for the race.

Jumping

Headwinds and tailwinds can have disastrous effects on horizontal jumpers and pole vaulters, since headwinds tend to shorten runup strides and tailwinds tend to lengthen them. Even the most experienced jumpers have trouble making step adjustments with consistency when the wind is blowing, particularly when there are occasional gusts.

In the vault, headwinds and tailwinds create additional problems because of the contribution of horizontal velocity to the bending of the pole. (See Chapter 18.) Velocity is the chief contributor to kinetic energy ($\frac{1}{2}mv^2$), which means that an increase or decrease in takeoff velocity will have an effect on handhold height, and could even require selecting a different pole.

For the most part, vaulters enjoy a tailwind because it increases the horizontal velocity of the approach run and thus enables them to hold higher on the pole. As long as the vaulter can make the necessary step adjustments for the approach run and has a strong enough pole, a tailwind is a definite aid.

Throwing

The effect of air resistance on shots and hammers in flight is usually ignored during analysis, since the resistance of air has little effect on their flight curves. However, air resistance becomes a very important factor when considering the flight curves of discuses and javelins because of the aerodynamic qualities of these two implements. (A detailed explanation of the effects of air resistance on discuses and javelins follows in Chapters 23 and 25.)

3
CENTER OF MASS

The mass of a body is the amount of material of which it is made. (Mass is not the same as weight. Weight is the force of attraction between a body and the earth.)

The center of mass (or center of gravity) of a body is the point where all of its mass may be considered to be concentrated. The center of mass of a 16-lb. shot is near its center, of course, while a doughnut's center of mass is near its center, but it is also in space, demonstrating that the center of mass of an object does not necessarily have to be within the substance of the object.

In the human body, the center of mass is not a fixed point located in a specific part of the body. As body position changes, the center of mass also changes location.

In a standing position, a person's center of mass is approximately 1-2" below the navel. (This depends, of course, on the person's weight, body shape, etc.) By raising both arms straight up, the center of mass is raised about 2½". Bringing one arm down and to the side lowers the center of mass and moves it to that side of the body. By bending at the waist, the center of mass may be in space, outside the body. (See Figure 4.)

There are a number of track & field events in which the body's center of mass is outside the body at some time during the execution of the event. In long jumping, for

Figure **4**. **A** person's center of mass is not a fixed point. When the arms are raised, the center of mass raises. Extending an arm to one side shifts the center of mass to that side. Bending the body at the waist moves the center of mass into space.

example, the center of mass is within the body during most of the flight to the pit, but it shifts forward, just outside the body as the jumper assumes the landing position. (See Figure 5.)

Figure 5. The long jumper's center of mass shifts forward, into space, just before the landing.

Stability And Balance

A body's stability when at rest is dependent upon the weight of the body, the height of the body's center of mass, the size of the body's base of support, and most important, the relationship of the center of mass to the base of support. For the body at rest to be in balance, its center of mass must be above at least a portion of the base of support.* If the center of mass is not above the base, either the base must be adjusted or the body must topple. (See Figure 6.)

A person can give the illusion of standing "out of balance" by leaning to the side and using the opposite leg to expand the base of support. However, a line drawn downward from the center of mass shows that it is actually above a portion of the base of support, even though it appears to the eye that it is not. (See Figure 7.)

When getting up from a chair, the base of support shifts from the person's seat to the feet. This requires bending forward before rising, shifting the center of mass into a position above the feet. (See Figure 8.)

Without bending forward, getting up is not possible. However, it is possible to get up without bending forward if the person holds heavy weights in his outstretched hands. (See Figure 9.) Because the heavy weights move the center of mass (of the person and the weights combined) forward, above the feet, the person is able to stand up without bending forward.

Balance And Motion

If a person standing in a state of balance leans forward slowly until the body's center of mass is no longer above the feet, the natural tendency is to step forward with one foot. If the leaning continues, another forward step (or a series of steps) will keep the person from falling forward. Of course, this continual adjustment of the base of support is called walking, which is performed as a reflex action, although it is actually learned during infancy with some difficulty.

If the center of mass is continually shifted even farther ahead of the base of support, then the person must increase

* It is presumed that no artificial supports are being used, such as nailing the person's shoes to the floor.

the frequency of the steps and change from a walking technique to running. Again, this is a process which is learned at an early age but seems completely natural later in life.

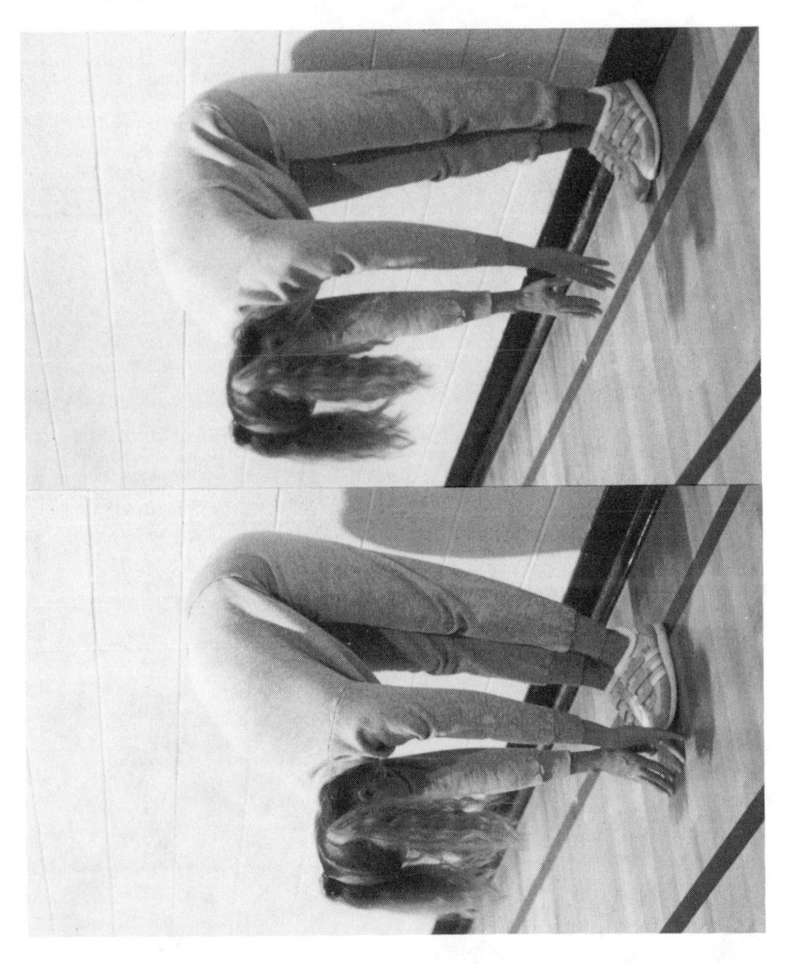

Figure 6. Bending forward to touch the toes can only be done as long as the body's center of mass is above the feet. As the person bends forward, the buttocks must shift backwards for the body to remain in balance. If the same movement is attempted while the feet are against a wall (which restricts the buttocks from shifting backwards), the center of mass can no longer be above the base of support and the person must fall forward.

Figure 7. The illustration of leaning against an imaginary object is possible because the subject's right foot provides the edge of his base of support. As long as the center of mass is above any portion of the total base of support, a person can remain in balance.

Figure 8. Getting up from a chair requires bending forward, shifting the center of mass into a position above the feet.

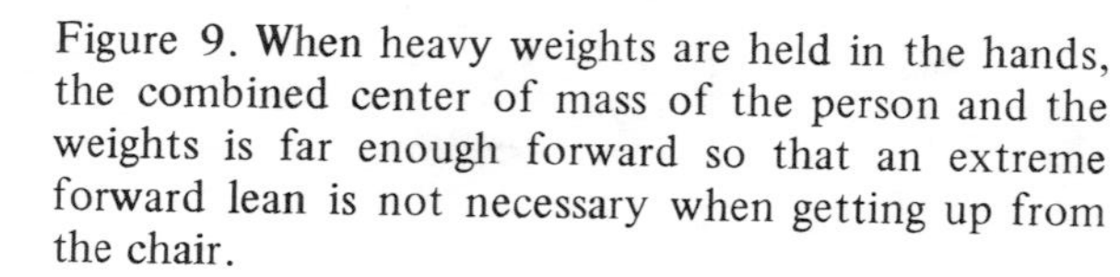

Figure 9. When heavy weights are held in the hands, the combined center of mass of the person and the weights is far enough forward so that an extreme forward lean is not necessary when getting up from the chair.

Figure 10. In the "set" position, the sprinter's center of mass is above a wide base of support.

Sprint starting requires a sudden change in the relationship of the sprinter's center of mass to the base of support. In the "set" position, the center of mass is over the forward portion of a wide base of support—the runner's hands behind the starting line and the feet in the blocks. (See Figure 10.) But as the gun fires, the sprinter lifts his hands from the track and the base of support—now only the feet in the blocks—is suddenly well behind the center of mass. (See Figure 11.) Only great acceleration from the blocks keeps the sprinter from falling forward or from stumbling while coming out of the blocks.

Figure 11. When the gun is fired, there is a sudden change in the relationship of the sprinter's center of mass to the base of support.

4
CURVES OF FLIGHT

As was mentioned in Chapter 1, an athlete or throwing implement leaves the ground or thrower's hand at an angle and speed that is determined by two separate velocities—one horizontal and one vertical. After the athlete or implement is free in the air, its center of mass follows a specific flight curve—a parabolic curve for humans, shots, and hammers; an aerodynamic curve for discuses and javelins.

Parabolic Curves

The moment a body (a human body or an inanimate object, such as a shot or hammer) leaves the ground, the entire flight path of its center of mass is determined. The combination of forward-upward velocity when it leaves the ground and the force of gravity during the flight in the air causes the body's center of mass to follow a perfectly regular curve called a parabola, or parabolic curve. The horizontal component of the takeoff or release velocity is unaffected by outside forces (except for some air resistance, which is of little significance), but gravity gradually slows the vertical component to zero and then reverses the process, causing the body to fall at the same angle and velocity as it left the ground. The range (distance from takeoff or release to landing) of the parabolic curve is determined by horizontal velocity and the time of flight; the height is determined by vertical velocity.

No human movements can alter the flight path of an athlete's center of mass once he is airborne. Changes in body position can change the position of the center of mass within the body during the time in the air, but wherever the center of mass happens to be in relation to the body, and no matter how much it changes position within the body, the center of mass continues to follow the parabolic curve that was established at takeoff.

In high jumping, it is obvious that the jumper's center of mass must be projected high enough at takeoff to get the entire body over the bar, and that the peak of the parabola be over the bar. No matter how efficient the jumper's bar clearance style might be, it is worthless if the takeoff has not been good.

Some high jump layout styles allow the jumper's center of mass to pass closer to the bar than others. The important

Figure 12. The high jumper who keeps the center of mass low during bar clearance does not have to jump quite as high to clear the height.

point in bar clearance is that as much body mass as possible be below the bar at the peak of the jump, so that the center of mass need not be projected quite so high at takeoff in order to clear the height. (See Figure 12.)

In long jumping, sweeping the arms backward just before landing in the pit can add distance to the jump. With the arms held back, the jumper's center of mass moves backward in his body, and thus, the entire body position moves forward, since the center of mass is following its parabolic curve.

In hurdling, there is an advantage in lowering the head after leaving the ground. Lowering the head lowers the center of mass in the body, which means that the entire body is raised slightly as the center of mass follows the predetermined parabolic curve. The hurdler who lowers his head during hurdle clearance does not have to project his center of

Figure 13. In hurdling, lowering the head can reduce the amount of time required to clear the hurdles.

mass quite as high and does not have to spend quite as much time in the air. (See Figure 13.)

After the release, shots and hammers also follow parabolic curves, but discuses and javelins, which are noticeably affected by air resistance, do not.

Aerodynamic Curves

Throwing implements affected by air resistance (discuses and javelins) follow irregular flight curves after being released because of their aerodynamic qualities. These implements are designed to "catch the air," lengthening their time in the air (and their distance from release to landing).

An aerodynamic implement's angle of attack (the angle between its plane and the direction of the relative wind it is encountering during flight) determines the extent to which it is held up, prolonging its flight, and the extent to which its forward motion is slowed. The thrower must develop a technique which will allow an angle of attack at release that affords the combination of lift and drag which produces the best possible result.

Optimum Angles For Achieving Horizontal Distance

The optimum angle for the projection of a missile (to achieve the best horizontal result) is 45 degrees–*if the point of landing is at exactly the same level in altitude as the height from which the missile is projected.* Obviously, this is never the case in the track & field events. Throwing implements and horizontal jumpers' centers of mass are *above* the ground level when flight begins, but the results are measured *at* ground level, requiring release and takeoff angles which are less than 45 degrees.

An experiment with water from a garden hose demonstrates the effects of various angles of projection on horizontal distance. If you hold the end of the hose near ground level and raise and lower the angle of the nozzle until the angle is established that allows the stream of water to land the greatest distance from the nozzle, you will find the angle of projection to be approximately 45 degrees. (See Figure 14.) Angles that are less or greater than 45 degrees cause the water to fall short of the maximum distance.

If you raise the height of the nozzle and again adjust the angle so that the greatest possible distance is covered by the stream of water, two things become apparent regarding the relationship of height of release or takeoff and the projection of a missile. (See Figure 15.)

1. The optimum angle for the projection of a missile lowers as the height of release or takeoff is raised.

2. Increasing the height of release or takeoff automatically increases the distance a missile is projected. (See Figure 16.) This point is not made in an effort to improve a particular athlete's performance, since the improvement would require an increase in the athlete's height. It is made to show that taller athletes have a decided advantage over their shorter opponents. In the shot put, for example, a one-foot increase in height of release can increase the distance of the

Figure 14. When a hose nozzle is held at ground level, the optimum angle for projecting a stream of water for maximal distance is 45 degrees.

put by 9-15", depending upon the angle and velocity of release. (See Chapter 22.)

Throwing Implements

The optimum angle for the projection of a throwing implement depends upon the height of release, the velocity of release, and, in discus and javelin throwing, on the angle of attack and the aerodynamic properties of the implement and, *especially,* the angle of attack at which it is released.

In shot putting, the optimum release angle is between 40 and 42 degrees. The angle can be plotted by bisecting the angle formed by a line drawn vertically upward through the shot at release and a line drawn from the shot to the eventual landing point.

Because of aerodynamic forces, the discus and javelin do not follow parabolic curves in flight. For this reason, the release angles are even lower than in the "parabolic throws"

Figure 15. When the hose nozzle is held above ground level, the maximal distance, which is increased as the nozzle is raised, can only be reached by lowering the angle of the nozzle.

(shot and hammer). The optimum release angle for the discus is between 35 and 40 degrees and for the javelin it is between 30 and 35 degrees.

Horizontal Jumpers

In long and triple jumping, the athlete develops great horizontal velocity during the approach run, making it difficult to obtain vertical velocity at takeoff. The approach is so fast that nothing can be done at takeoff to add vertical velocity that compares with the horizontal velocity already established, so the takeoff angle must be considerably below 45 degrees. In fact, it is seldom above 25 degrees. Higher takeoff angles can be achieved by slowing down during the approach run or at takeoff but the result is always a shorter jump.

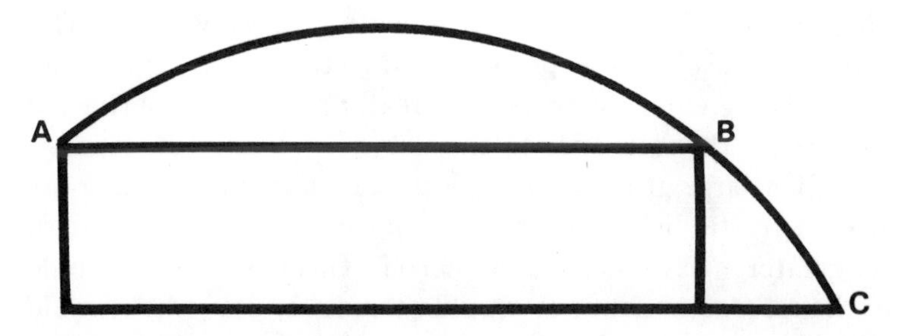

Figure 16. In this simple illustration, if A and B are at ground level, an object projected from A would follow a parabolic curve and land at B. If A is *above* ground level and C is at ground level, the object projected from A would follow a parabolic curve and land at C. The height of release has increased the horizontal distance automatically.

5
INERTIA AND MOMENTUM

Inertia, the Latin word for inactivity or laziness, is *resistance to change in motion.* A body at rest tends to remain at rest, resisting movement until a force causes it to move. A moving body tends to continue moving, resisting a change until forces cause it to speed up or slow down. This resistance to change is proportional to the mass of the object itself. The greater the mass, the greater the body's inertia and the greater the resistance to change.

If a large truck and a sports car decided to race away from a traffic light, the truck would lose the race because of its greater mass and greater inertia. The two vehicles could exchange engines, but it wouldn't matter. The truck would still lose the race. If the two vehicles were traveling along open highway at the same high speed and the drivers both slammed on their brakes, it would take the truck a much greater distance to stop, again because of its greater inertia.

The inertia of 16-lb. shot is twice that of an 8-pounder. It is because of this difference in inertia that a 16-lb. shot cannot be put as far as an 8-lb. shot.

If two sprinters, one weighing 100 lbs. and the other 200lbs., are in the starting blocks side by side when the gun is fired, the 200-pounder must exert twice the force exerted by the lighter sprinter, just to stay even. If they both exert the same force, the 100-pounder will accelerate at twice the rate of the heavier sprinter.

If the 200-pounder could reduce his weight to 180 without reducing leg strength he would he 11% faster out of the blocks than when he weighed 200. In other words, in order to get faster starts a sprinter must develop strong legs while keeping body weight as low as possible. (Once in full stride, however, it is slightly easier for the heavier sprinter to maintain velocity because of inertia and it takes the heavier sprinter longer to stop after crossing the finish line.)

Momentum is *mass x velocity.* Everything that moves has a certain amount of momentum. If either the mass of the body or its velocity is increased, without decreasing the other factor, the body's momentum must increase. Conversely, a decrease in either factor produces a decrease in momentum.

Every time an athlete or a throwing implement moves, momentum is developed. A 100-lb. sprinter running 16 fps (less than half the velocity of a world-class sprinter) has 1600 units (100 x 16) of momentum. A shot weighing 16 lbs. traveling 100 fps (more than twice a normal putting velocity) also has 1600 units of momentum. However, neither the sprinter nor the shot has as much momentum as a bullet as it leaves the barrel of a rifle. Even though the bullet has considerably less mass, the high velocity of the bullet produces a greater momentum.

In sports movement, there is no opportunity to increase or decrease the mass of an athlete (or throwing implement) during the time of a particular movement. Therefore, momentum must be directly linked to the velocity of the body in motion. An increase in velocity increases momentum. A decrease in velocity decreases momentum.

An understanding of momentum is more important in sports involving collisions (such as football) than in track & field events. However, it is necessary to include momentum among the biomechanical concepts here as a basis for understanding the concept of *rotary momentum* later.

6
GROUND REACTION

An athlete is able to run, jump, and throw because of ground reaction forces—forces equal and opposite to those forces applied by the athlete against the ground.* The greater the forces appled to the ground by the athlete, the greater the forces given back to the athlete by the ground.

If an athlete who weights 160lb stands on the ground, the ground reacts by pushing back with a force of exactly 160lb. If the athlete suddenly pushes against the ground with a force of 200lb, the ground pushes back with an equal force of 200lb and the athlete jumps into the air (or lifts a throwing implement) with 40lb of lift. The greater the effective force (the total force minus the athlete's body weight), the greater the force the ground will return, and the greater the lift. (See Figure 17.)

The force against the ground (thus, the force returned to the athlete) can be increased with an increase in takeoff-leg strength, and by initiating limb movements away from the ground during the time the ground forces are applied. (See Figure 18.)

* Although the word *ground* usually refers to the earth's surface, it can also refer to objects above the earth's surface that are in contact with the ground. A pole vaulter, for example, is in contact with the ground until he releases the pole.

Figure 17. When forces are exerted against the ground by an athlete, the forces are returned to the athlete by the ground.

Figure 18. Limb movements away from the ground increase the forces the athlete is able to exert against the ground.

Impulse

It is not only the amount of force applied to the ground that determines the velocity with which an athlete leaves the ground (or an implement leaves the athlete's hand), but also the *time* during which that force is applied. (Force x Time = Impulse.)

A force acting for a long time can produce the same

impulse as a greater force acting for a shorter time. The time required (or available) for applying the force determines the combination of force and time that will produce the greatest impulse and the potential for the best results. In the throwing events, for example, the forces are applied to the implements over the greatest possible distance (and time) in order to increase impulse. On the other hand, the high jumper's impulse can be increased greatly by applying the force against the ground for the shortest possible time. (Bending the lead leg at the knee and the arms at the elbows shortens the high jumper's time on the ground.) The increase in force offsets the decrease in time enough to produce a greater impulse and, thus, greater vertical velocity at takeoff.

7
ROTARY MOTION

Rotary motion, also called *rotation,* is turning or rotating motion in a circle (or arc) around an axis. All of the mass outside the axis is in motion around the axis, while the axis itself remains in a fixed position relative to the motion around it.

Axes

Axes are invisible straight lines passing through turning bodies. Anything that turns, whether it is on the ground or in the air, turns around at least one axis and thus has rotary motion. All parts of the body outside that axis turn at right angles to the axis.

When an athlete's body is rotating in a vertical plane while in contact with the ground, such as when a javelin is released, the axis passes horizontally through the point where the foot meets the ground as the foot is planted as the upper body rotates forward. (See Figure 19.)

When the athlete is in contact with the ground and the rotation is in a horizontal plane, such as in the discus turn, the axis passes through the point of support and the athlete's center of mass. (See Figure 20.)

Primary Axes

When an athlete leaves the ground and is free in the air,

Figure 19. The javelin thrower rotates forward during release, over an axis that passes through the point where the lead foot is in contact with the ground.

Figure 20. The discus thrower's axis of rotation passes through the thrower's point of support and his center of mass.

all turning of the entire body must be around primary axes that pass through the body's center of mass. The three primary axes, which are mutually perpendicular, are the longitudinal (or long) axis, which runs from head to toe; the transverse axis, which runs from side to side; and the frontal axis, which runs from front to back. (See Figure 21.)

Any object spinning in the air must turn around one or more of the primary axes which pass through its center of mass. For example, try spinning a carpenter's hammer in the air, throwing it and catching it by the handle. Because the hammer's center of mass is in the head, far from the center of the hammer, the handle rotates around the head as it turns in the air.

Nutation

When rotation in the air is around two primary axes at the same time, the object nutates, or wobbles (like a spinning top). The most obvious example of nutation in track & field is a discus which has been thrown incorrectly, with the force from the hand applied across—rather than directly along—the plane of the discus at release. The result is a wobbling discus in flight.

Secondary Axes

When an athlete is airborne and wishes to temporarily slow or reverse unwanted body rotation, or wants to change his body position before landing, he may do so by creating movement around a secondary axis—an axis which does *not* pass through his center of mass.

For example, a downhill skier who is off balance in the air swings his arms (and sometimes even his ski poles), windmill fashion, to regain his balance before landing. The secondary axis, created through the skier's shoulders, helps him change his body position in the air.

Secondary axes may be employed while in contact with the ground, too:

Figure 21. The three primary axes are longitudinal (head to toe), transverse (side to side), and frontal (front to back). All primary axes must pass through the body's center of mass.

The beginning roller skater swings his arms, sometimes violently, to regain his balance. The circus wire walker (if he doesn't carry a balance pole) swings his arms to stay balanced on the wire.

The sprinter who bends forward and leans into the tape in a close finish may have to swing his arms, windmill fashion, in order to regain his balance and keep from falling headfirst onto the track.

The long jumper who hitch-kicks in the air may not realize how important that movement is—or that he is actually creating *two* secondary axes when he "runs in the air." By creating these secondary axes, the jumper is able to delay forward rotation, enabling him to assume a better landing position.

8
ROTARY INERTIA AND ROTARY MOMENTUM

Inertia, as was explained in Chapter 5, is resistance to change in motion, and is proportional to the mass involved. The greater the mass, the greater the inertia.

Rotary inertia is resistance to change in rotary motion (around an axis). However, with rotary inertia the resistance to change is dependent not only upon the mass, but also upon the distribution of that mass around an axis. The farther the mass is from the axis of rotation, the greater the resistance to speeding up or slowing down. The closer the mass is to the axis, the less the resistance.

A sprinter's arms swing forward and backward around an axis through the shoulders. The arms swing more easily and more rapidly if they are bent at the elbows than if they are straight, since the bent arms distribute the mass of the arms closer to the axis of rotation, which reduces rotary inertia. Likewise, the sprinter's recovery leg (which rotates around an axis through the hips) comes forward with greater speed if the knee is bent as much as possible. (See Figure 22.)

High jumpers and horizontal jumpers spend less time on the ground at takeoff when the leading leg is bent at the knee and both arms are bent at the elbows. (See Figure 23.) Lengthening any of these limbs increases rotary inertia, which increases the time required to swing them forward and upward. This in turn, slows the takeoff.

Figure 22. To insure fast limb movements, a sprinter's arms should be bent at the elbows and his leg should be bent at the knee as much as possible during the recovery stride.

Figure 23. Bending the knee produces a faster takeoff.

The pole vaulter's swing-up on the pole can be easier and faster when he bends one or both legs at the knee as he swings forward. By bringing his mass closer to the axis of rotation (a point where his top hand is gripping the pole), rotary inertia is reduced.

Rotary Momentum

Rotary momentum is the *turning* momentum (around one or more primary axes) of a body. It depends on the rotary velocity of the body and the body's rotary inertia. (Rotary Momentum = Rotary Velocity x Rotary Inertia.)

A turning body, once it is airborne, has constant rotary momentum which had to begin on the ground and cannot be increased or decreased once the body is in the air. Rotary momentum must remain constant until the body returns to the ground.

Conservation Of Rotary Momentum

Since rotary momentum cannot be changed once an athlete is in the air, and since an athlete's rotary momentum is the product of two factors–rotary velocity and rotary inertia–then, obviously, reducing either of the factors must automatically increase the other. A decrease in rotary inertia (by pulling the mass closer to the axis) increases rotary velocity. An increase in rotary inertia decreases rotary velocity. This concept is called conservation of rotary momentum.

Conservation can be demonstrated by having someone stand on a "frictionless" turntable while holding a heavy object (such as a 12-lb. shot) in each hand. (Standing on a turntable simulates being in the air and free to turn around the body's longitudinal axis.)

Have the subject hold the shots at arms' length. (See Figure 24.) Then, turn him very slowly around his longitudinal axis, which establishes his rotary momentum. As soon as he is turning on his own, have him reduce rotary inertia by

Figure 24. Conservation of rotary momentum can be demonstrated with the help of a turntable and two shots.

pulling the shots in toward his body. The velocity of turning immediately increases. As soon as the shots are returned to their original positions, which increases rotary inertia, the turntable slows to its original rotary velocity (minus only losses incurred through friction). The more mass (the heavier the objects) held away from the body when the turning motion begins, the greater the increase in velocity when the mass is pulled in toward the axis of rotation.

A long jumper experiences unwanted rotary momentum (forward rotation) around the transverse axis while airborne. If the jumper assumes a tucked position (such as in the "sail" technique) rather than an outstretched position, the velocity of turning will increase markedly because of the reduced rotary inertia and his chances of assuming an efficient landing position will be greatly reduced.

A pole vaulter can slow the turning of his body around the transverse axis after he clears the crossbar by raising his arms over his head as he falls toward the pit. (See Figure 25.)

Discounting friction for the moment, conservation of rotary momentum can also take place on the ground, as is the case with a figure skater who is able to spin so rapidly on the ice that the movements seem a blur to the human eye. The skater accomplishes this by keeping arms and legs wide as the spinning movement begins, then pulling the limbs in close to the axis of turning, greatly increasing rotary velocity.

The same principle may be used in discus throwing. By starting the turn with a wide base, with arms and legs as far from the longitudinal axis as possible, and then reducing rotary inertia before the release, the velocity of the release may be increased greatly. (See Figure 26.)

In shot putting, discus throwing and javelin throwing, velocity of release may be increased by bending the lead arm at the elbow (or better, by bringing the entire arm in close to the body) during the release action. Bringing the arm in reduces inertia and increases the upper body's turning velocity.

Figure 25. When the pole vaulter is falling toward the pit, raising the arms increases rotary inertia around the transverse axis, which reduces rotary velocity.

Figure 26. Rotary inertia around the longitudinal axis is great when the discus turn begins. Reducing rotary inertia at the time of release increases rotary velocity.

9
ROTATION IN THE AIR

When an athlete leaves the ground, there is usually at least some rotation of the athlete's body or of parts of the body while the athlete is airborne.

If the rotation is initiated while the athlete is still in contact with the ground, the rotation in the air continues around an *axis of momentum* until the athlete returns to the ground.

If the rotation is initiated while the athlete is in the air, the rotation is around an *axis of movement,* and the rotation creates an immediate equal and opposite reaction of another part of the body around the same axis.

Rotation Originating On The Ground

In most sports movements, athletes leave the ground, at least momentarily, and turn briefly around one or more of the three primary axes while they are in the air. An athlete's rotary momentum, which must come from the ground, is the result of various ground reaction forces at or prior to take-off. When the resultant force is "off center," that is, when it does not pass directly through the body's center of mass, rotation is begun and it continues in the air. This off-center line of force is called *eccentric thrust.*

Vertical eccentric thrust. When two basketball players are involved in a jump-ball situation, they crouch, jump up in the air, reach for the ball with their fingertips, and return to the playing floor. This is one of the rare cases in sport in which athletes leave the ground, are free in the air, and return to the ground in the same spot, with no rotation. Each basketball player's vertical thrust from the ground is directly through his center of mass and, thus, the force is *not* eccentric.

Eccentric thrust, on the other hand, creates rotary motion, but it does so at the expense of effective force. The more eccentric thrust there is at takeoff (and, thus, the more rotary momentum around one or more axes), the less effective force (and, thus, the less the height or distance attained).

Using a relay baton, you can demonstrate the effects of vertical eccentric thrust on height and rotation. First, balance the baton horizontally on the side of your forefinger and toss it straight up a foot or two. If the force from the finger is directly through the baton's center of mass, it will achieve a certain height and will have no rotation. Then, holding the baton in the same horizontal position with the opposite hand, use the same finger to toss the baton up with the same force, but this time apply the force a distance from the center of the baton instead of through its center of mass. Rotation is immediately evident and the height of the throw is decreased. The farther from the center the force is applied, the greater the rotation and the less height achieved.

In high jumping, a certain amount of rotation is necessary, of course, in order for the jumper to achieve an efficient layout position and to rotate over the crossbar. But because increases in rotation automatically reduce the jumper's ability to achieve height, high jumping becomes a compromise. The jumper must get enough rotation for bar clearance, but not any more than is absolutely necessary. (It should be emphasized that the technique of thrusting eccentrically is developed by the jumper through trial and error in practice sessions. He does not realize that he is doing it, nor is he likely to know it is curtailing his "height-getting ability.")

Horizontal eccentric thrust. When the linear motion (motion along a straight line) of a rigid object is interrupted at one end of the object, the other end of the object continues ahead, but at an increased speed. If the object leaves the ground following this "interruption," forward rotation is begun because of horizontal eccentric thrust.

In track & field, horizontal eccentric thrust takes place when an athlete is running and a foot is "planted," allowing the upper body to continue moving forward. The axis of rotation is the point where the foot meets the ground. (See Figure 27.)

The long jumper's horizontal eccentric thrust at takeoff causes his body to acquire unwanted forward rotation in the air, which may then be neutralized through employment of secondary axes (page 43).

The javelin thrower plants his foot prior to releasing the javelin, causing his upper body and arm (and, of course, the javelin) to increase in velocity for the release.

Figure 27. The long jumper's upper body rotates forward as soon as the takeoff foot has been planted.

Rotation Originating in the Air

When an athlete attempts to turn his body while in the air without first obtaining some rotation from the ground, all he can do is initiate an action, which produces an equal and opposite reaction. The action and simultaneous reaction can either be in the same plane or in planes that are parallel. (See Figure 28.)

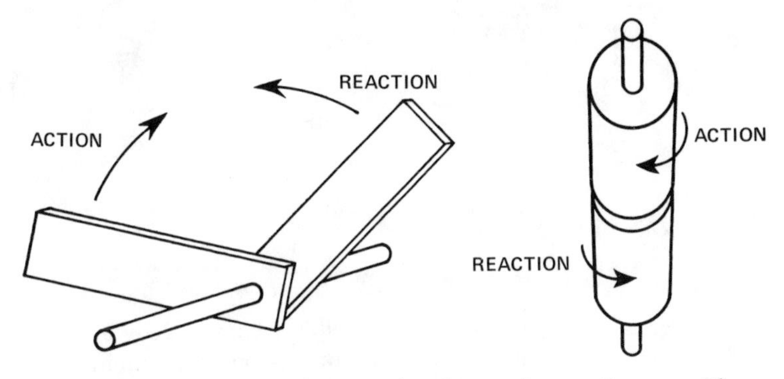

Figure 28. An action and its equal and opposite reaction can either be in the same plane (left) or in planes that are parallel (right).

Action-reaction in the same plane. In the final stages of the long jumper's flight, the legs are extended forward in preparation for the landing, and the trunk and head are also bent forward. If the jumper then decides to raise his upper body (or his head so that he can see better), the legs must lower in reaction, reducing the length of the jump. (See Figure 29.) The lower the jumper can keep the upper body, the higher the legs will remain in reaction. This is an example of action-reaction in the same plane (often called "jack-knifing" in coaching circles).

In high jump clearances, the legs are brought up to clear the crossbar, and the equal and opposite reaction in the upper body is the maintaining of the relative position of the head and shoulders, which have been rotating toward the pit. (See Figure 30.) The transverse axis (side to side) is still an

Figure 29. When a long jumper raises the upper body or head before landing, the legs must drop in reaction.

axis of momentum with rotation from the ground, but it also has become an axis of movement, with action on one side of it and reaction on the other side.

In hurdle races, the runner leans his trunk forward as he clears each barrier, but he can only do this if he also raises his lead leg, for the two movements cannot be separated. One movement (which one doesn't really matter) is the action and the other is the equal and opposite reaction. Then, as the trailing leg clears the hurdle, a sudden lifting of the upper body snaps the lead leg back toward the ground in reaction.

Similarly, the pole vaulter raises his arms when clearing the crossbar. His legs react by moving backwards and up. (See Figure 31.)

Action-reaction in parallel planes. A simple demonstration of action-reaction in parallel planes can be performed

Figure 30. The high jumper raises the legs when clearing the crossbar, an action which prevents the jumper's body from rotating into a head-down position.

Figure 31. The reaction to the pole vaulter's raising of the arms when clearing the crossbar is an equal and opposite backward and upward movement of the legs.

by jumping straight up in the air without rotating and at the peak of the jump, swinging your arms in one direction around your longitudinal axis. Your legs, in reaction, will immediately rotate in the opposite direction in a horizontal plane that is parallel to that of your arms. If this example is not graphic enough, do it again, this time jumping in the air and throwing a heavy object to one side. (See Figure 32.)

For even more complete experimentation, stand on a turntable or twisting board (which simulates being free in the air around the longitudinal axis) and turn the upper body in one direction. The lower body *must* turn in the opposite direction. (See Figure 33.)

The action of the hurdler's forward-moving trailing leg and the movements of the arms in reaction is a good example

Figure 32. Jumping in the air and throwing a heavy object to one side brings about an equal and opposite rotary reaction in the lower half of the body.

of action-reaction in parallel planes. (See Figure 34.) As the hurdler's trailing leg comes forward over the hurdle, the equal and opposite reaction is the backward movement of the lead arm. If the hurdler is to remain facing straight ahead throughout his flight over the barrier, the action and the reaction must be equal. But since the leg has much more mass than the arm, the arm must travel through a wider arc than the leg and swing out wider (sometimes even into the next lane) in order to keep the upper and lower parts of the body in balance.

Figure 22 Action reaction in parallel planes can be demonstrated on a

Figure 34. The reaction to a hurdler's forward-moving trail leg is the backward-sweeping lead arm.

10 CENTRIPETAL AND CENTRIFUGAL FORCES

Centripetal force is a force that pulls *toward* an axis of rotation. Centrifugal force is a force that pulls *away from* an axis of rotation. The two forces are equal and opposite.

A sprinter running around a curve experiences an inward centripetal force against his feet which allows him to change direction (to follow the curve) while an equal and opposite centrifugal force is applied outward by the feet. To combat these forces, the runner must lean toward the inside of the curve. (See Figure 35.) The tighter the curve, the greater the centripetal force against the feet and the greater the inward lean in reaction.

Whenever the forces a runner applies to a running surface are not perpendicular to that surface (as when leaning inward on a curve), there is an automatic reduction in the runner's speed. The greater the lean, the greater the reduction, which means that the slowing effect of curve running is slightly greater in the inside lanes than in the outside. (Of course, some of the slowing effect created by leaning inward can be reduced if the track's curves are banked.)

The high jumper's curved approach run produces a lean away from the bar, the result of centripetal and centrifugal forces. (See Figure 36.) This lean insures a more vertical posture and an increase in ground reaction forces at takeoff. (See Chapter 17.)

Figure 35. Curve running requires an inward lean, the result of centripetal and centrifugal forces.

Figure 36. The high jumper's curved approach produces a lean away from the bar.

Among the many examples of centripetal and centrifugal forces in the field events, two of the most obvious are in hammer and discus throwing. In each event, the throwers initiate rotary velocity around a vertical axis, applying centripetal force to the implement, pulling it toward the axis of rotation, while the implement is exerting centrifugal force on the thrower. (See Figure 37.) As rotary velocity around the vertical axis increases, the centripetal and centrifugal forces increase. As soon as the implement is released, it travels in a straight line, at a right angle to the turning radius.

Figure 37. The hammer thrower experiences centripetal (toward the axis of rotation) and centrifugal (away from the axis of rotation) forces before the hammer is released.

PART II

THE RUNNING EVENTS

11
BASIC BIOMECHANICS OF RUNNING

The running events can be divided into two general categories–the primarily anaerobic (oxygen independent) events, such as the sprints, hurdles, and sprint relay races, and the primarily aerobic (oxygen dependent) events, such as the middle distance and long distance races.

The Running Stride

Human runnning permits the body to float in the air between strides, with both feet off the ground approximately half the time. Thus, the runner's strides can be considerably longer than the length of the legs would otherwise allow. There is never a time when both feet are on the ground, as is the case with walking.*

In steady running, each running stride can be divided into three distinct parts, regardless of the speed:

* Race walking, which is not covered in this book, requires completely different techniques from those required for running. (Although this author has always been neutral on the subject, there are some coaches who believe race walking should not be considered a part of any track and field program. They say that determining who can walk the fastest is like determining who can whisper the loudest.)

Figure 38. The driving phase takes place while the center of mass is ahead of the foot that is on the ground.

1. *Driving phase.* The body is pushed forward by extending the hip, knee, and ankle joints of the driving leg while it is behind the body's center of mass. The driving phase continues until the driving foot leaves the ground. (See Figure 38.)

2. *Recovery phase.* Both feet are in the air as the driving foot leaves the ground well behind the body's center of mass. (See Figure 39.)

3. *Braking phase.* The opposite foot touches the ground a little ahead of the center of mass, causing a brief braking action. The body moves forward until the center of mass has moved ahead of the foot that is in contact with the ground. (See Figure 40.) Then the next driving phase begins.

The distance covered with each complete running stride

Figure 39. The recovery phase takes place while the center of mass is ahead of the foot that is on the ground.

Figure 40. The braking phase takes place from the time the foot touches the ground until the center of mass moves ahead of it.

is called the *stride length.* The number of strides taken in a given time is called the *stride frequency* or *cadence.*

Stride Length And Stride Frequency

Success in any running event is dependent upon two factors–stride length and stride frequency. The product of these two factors–the distance covered with each stride and the number of strides taken in a given time–equals the runner's speed.

In theory, an increase in either stride length or stride frequency will increase a runner's speed. However, each of these factors has such an effect on the other that there are times when increasing one reduces the other enough to produce a slower speed.

Stride length. It can be said that stride length is a function of running speed, since stride lengths tend to increase with increases in speed over the ground. However, a runner's *individual* stride length is determined by his leg length and by the ground reaction forces returned to the runner during each stride.

The length of the stride is measured to the point where the foot touches the ground during the recovery phase. If the foot touches too far forward (overstriding), there is a "braking" effect on the runner's stride, resulting in deceleration. If the foot touches too far back (understriding), stride frequency increases, but overall speed decreases. The "natural" stride in steady running is one in which the toe of the foot lands approximately 12" ahead of the body's center of mass during the recovery phase.

There have been attempts over the years to increase runners' natural strides unnaturally by forcing them to overstride during training sessions. But these methods produce only temporary results. The *natural* way for a runner to increase stride length is for him to increase the force against the ground in each driving phase. This, of course, requires increased leg strength. The resulting reaction from the ground drives the body's center of mass farther forward, lengthening the stride naturally, and the foot is still able to land approximately 12" ahead of the center of mass.

Stride frequency. It has long been thought that stride frequency is controlled by the runner's ability to contract and relax muscles, but we now know that there are other, more important considerations.

First, there is the direct relationship between leg length and both stride length and stride frequency. A sprinter with short legs has a naturally shorter stride, which brings the foot back to the ground sooner than if the stride were longer. Generally, the shorter the leg, he shorter the stride and the slower the frequency.

Second, it has been shown that runners have the ability to move their legs much faster during other activities (such as cycling) than they do when running, even when they are sprinting at their fastest speeds, but their stride frequency is limited *entirely* by the length of the stride. In other words, while the force against the ground during the driving phase determines the length of the stride, the resulting stride frequency is merely the time required to complete that stride. Forcing a greater frequency would produce a shorter stride and a reduction in speed. Obviously, then, the sprinter's training emphasis must be on improving stride length (without overstriding), not on stride frequency.

For many years, coaches have been devising training methods designed to increase stride frequency by forcing the runners to run faster than their current abilities would normally allow. Among them are the "tow method," in which athletes are pulled behind vehicles, and "running downhill using maximum stride frequency."

In the tow method, the tow provides a force that increases the runner's horizontal velocity. This may produce increased leg cadence, but it appears there would be an even greater increase in stride length than in stride frequency because of the increased horizontal velocity. In any case, as soon as towing stops, the runner's stride length and frequency return to normal.

Running downhill may increase the runner's speed over the ground (with the aid of gravity), but the chances of increasing stride frequency during this exercise are remote. When running downhill, the stride is lengthened considerably, just as the distance a shot can be put is increased with an increase in the height of release. (See page 33.) If the runner's point of takeoff height is higher than the point of landing, the strides are lengthened (and the frequency is reduced because of the additional time spent in the air), but *only while running downhill.* (See Figure 41.) As soon as the runner returns to a flat surface, stride length and frequency return to normal.

Uphill running, on the other hand, can be very beneficial for the runner. The angle of the ground surface to the body creates a simulation of the forward lean during acceleration, and the resistance of gravity contributes much more to the leg strengthening process than is accomplished in flat running.

Forward Lean in Running

The two factors which contribute to a runner's forward lean while running are his acceleration and, to a lesser extent, the amount of air resistance he encounters while running. Forward lean is the angle between the horizontal and a line from the runner's contact foot through his center of mass when the knees are side by side during the recovery phase. (See Figure 42.) An illusion of forward lean may be observed at the completion of the driving phase, when the runner is in the more classic running position. (See Figure 43.)

The greater the runner's acceleration, as in the early stages of a sprint event, the greater the forward lean. Backward lean, such as is observed in the closing stages of a sprint race, is the result of deceleration. After crossing the finish line, the sprinter decelerates more rapidly by swinging the free leg forward during the recovery phase of each stride

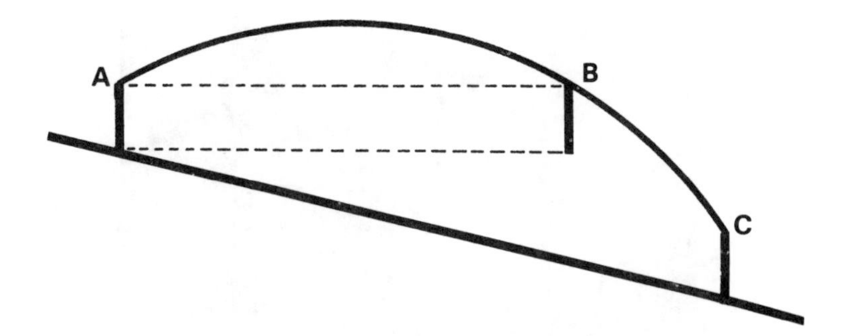

Figure 41. This simple illustration shows that downhill running requires an increased stride length and a decreased stride frequency. On flat ground, a runner's center of mass travels from A to B during one stride. While running downhill, the runner's center of mass travels from A to C, producing an unnatural (and temporary) increase in stride length and corresponding decrease in stride frequency.

Figure 42. A runner's lean is determined by drawing a line from his contact foot through his center of mass when his knees are side by side.

Figure 43. The "classic" running position gives the illusion of great forward body lean.

(overstriding), which produces a braking effect. When slowing down rapidly, backward lean is even more apparent.

Hurdling

Hurdling should be thought of as sprinting over barriers. The object is to run at top speed between barriers and to slow down as little as possible when negotiating each barrier, a technique which requires altering the sprinting technique with an exaggerated stride. (The distance between hurdles often prevents full-effort sprinting, since stride length is determined by the available distance.)

The higher the hurdle (and thus the greater the exaggeration of the hurdling stride), the more profound the effect on the running speed for the race. It may be surprising for some to learn that the increased hurdle height in the 110m race has a greater slowing effect on the runner than the endurance required to cover the 400m hurdle race.

The Olympic Record for the 110m hurdles (42" high) was set at an average speed of 8.33 mps, while the best ever for the Olympic 400m hurdles (36" high) is slightly faster–8.40 mps.

By comparison, the Olympic Record races for 100m and 400m without hurdles show that the 400 is run much slower because of the effects of endurance. The 100 race was covered at an average speed of 10.10 mps, while the 400-meter race was only 9.13 mps–almost 10% slower. This comparison further emphasizes the effects of hurdle height on performance and the need to develop the most efficient possible clearance technique in order to improve hurdling times.

12
SPRINTING

Success in sprinting, as in all running, is dependent upon stride length and stride frequency. Increasing either factor without an offsetting decrease in the other factor automatically increases a runner's sprinting speed.

Generally, there is an inverse relationship between stride length and stride frequency. Sprinters with long strides generally tend to have a slower cadence. Those with shorter strides have a faster cadence.

Improving Sprinting Speed

Although it has long been a belief among coaches that sprinters are born, not made, it has been proved repeatedly that this is not entirely true. It is true that certain physical qualities allow some athletes to run faster than others, but it is also true that anyone's sprinting ability can be improved though implementation of a regular program of leg strengthening exercises. Increased strength produces increased force against the ground, which produces an increase in natural stride length.

Analysis of slow motion movies and videotapes shows that most world-class sprinters of approximately the same height (and leg length) have about the same stride frequency. Sprinters often run in step in the stretch run to the tape; their stride frequencies are very close to equal. Thus, it is stride length that allows the faster sprinters to pull away from the field. Interestingly, even when one of the sprinters does have a greater stride frequency than the opponent (of the same approximate height), the longer-striding sprinter continues to pull away, obviously because of the stride length advantage.

From a training standpoint, stride length is far more important than stride frequency, but increasing stride length must be done in a natural way. In a 100m race in which a sprinter has an average stride length of 6'10", a 1" increase in stride length adds up to a 4' increase by the end of the race. However, if the sprinter had increased his stride length unnaturally by overstriding–running so each foot would strike the ground well ahead of the body's center of mass–stride frequency would have been reduced so much that his total speed for the race would have been reduced. Therefore, stride length must be increased in a natural way; the sprinter must be able to increase the force of each stride against the ground. And the way the sprinter can apply a greater force is by increasing leg strength.

Sprint Starting

Research has shown that "medium" block settings (those that neither crowd nor stretch the sprinter in the blocks) provide the faster starts. Ideally, optimal block settings allow the front knee to assume an angle of approximately 90 degrees and the back knee an angle of approximately 120 degrees when the sprinter is in the set position. (See Figure 44.)

Force against the blocks. A sprinter can improve his starting speed by increasing his leg strength, which in turn increases his force against the blocks. The greater the force the sprinter is able to apply to the blocks (particularly the front block), the greater the force the blocks will return to the sprinter, and the faster the start.

Figure 44. Medium block settings provide the fastest sprint starts.

If the sprinter's front leg is exceptionally strong, additional force can be applied to the block by assuming a set position in which the front knee is slightly less than 90 degrees. If the front leg is *not* particularly strong, the angle should be slightly greater than 90 degrees.

A sprinter's effective force against the blocks can be further increased by quickly driving the leading knee and the opposite arm forward at the sound of the gun, which increases the force against the front block and ensures an even faster start. (See Figure 45.)

Figure 45. Force against the front block can be increased greatly by quickly driving the leading knee and the opposite arm forward.

Body Mass

The sprinter's body mass has a decided effect on his ability to start fast. The greater the mass, the greater the sprinter's inertia (resistance to change in motion).

In order to get faster starts, a sprinter should try to develop strong legs, while keeping his body weight as low as possible. (Once he is in full stride, however, a sprinter's weight is no longer an important factor.)

Acceleration in Sprinting

As the sprinter leaves the blocks, his acceleration is very great at first. But as the ground begins to move faster and faster under the sprinter's feet, and as air resistance increases with each accelerating stride, he must reach a point when he cannot move his legs fast enough to keep his feet on the ground long enough to continue accelerating. At that point–about six seconds into the race–the sprinter's legs are moving so fast that the efficiency of the muscular contractions is considerably less than it was at the beginning.

The sprinter may maintain his top speed briefly, but because top-speed muscle contractions are possible for only a few strides (limited by muscular endurance), he begins a gradual, unnoticed deceleration, no matter how much he tries not to, all the way to the finish. It may appear to the spectators–and feel to the sprinters themselves–that there is continuous acceleration to the finish, but it simply cannot and does not happen. Everything else being equal, the sprinter who is in better condition "slows down the least" and wins the race.

There are coaches who claim that certain sprinters have the ability to accelerate at the finish of a sprint race. If these sprinters are running to the best of their abilities, then acceleration at the end of the race is impossible. Once the legs are moving at their fastest possible frequency, obviously it is not physically possible to increase that frequency, or to increase stride length without reducing that frequency.

Therefore, the only way to accelerate at the end of a sprint race is to deliberately decelerate at some earlier point in the race, or deliberately not accelerate to maximum speed from the beginning, either of which will contribute to running a slower total time for the race.

Acceleration and forward lean. Forward lean in running is a line between foot contact and the body's center of mass when the knees are closest together. It is not something that can be coached or learned. Forward lean is the direct result of the acceleration of the runner and, to a lesser degree, air resistance. The greater the acceleration, the greater the lean. (Acceleration is the cause and lean is the effect; never the opposite.)

If you were allowed to stand in the aisle of a modern jetliner during takeoff, you would find yourself leaning far forward during the period of the jet's greatest acceleration. Then as the plane reached its cruising speed, you would be standing straight up, with no lean at all. During the deceleration of landing, you would find yourself leaning backwards.

Except for the effects of air resistance (which you wouldn't have to consider inside the plane), the same situations are true in sprinting. During the period of greatest acceleration (coming out of the blocks), the sprinter has the greatest forward lean. (See Figure 46.) When acceleration decreases to zero, the only forward lean is the result of air resistance. In the closing stages of a sprint race (particularly noticeable in the 200 and 400), it is natural for the sprinter to have a slight backward lean. (See Figure 47.)

Thus, it is never necessary for a coach to worry about (or try to coach) forward lean. Sprinters may assume a more bent-over position, in an effort to please a coach who demands more lean out of the blocks, but actual lean remains the same. Except for the slight natural, uncontrollable lean against air resistance, forward lean can be increased by the athlete only by increasing his acceleration, and that is possible only for a short period of time.

Figure 46. A sprinter's forward lean is greatest early in the race when acceleration is the greatest.

Figure 47. Sprinters experience a backward lean during the closing stages of a race.

Air Resistance

The other factor contributing to forward lean is the resistance of the air against the sprinter. As speed increases, so does air resistance, and, to counteract the tendency to fall backwards, forward body lean must increase. When the sprinter faces a headwind, the lean is more acute. When there is a trailing wind, the posture is more upright.

As the sprinter accelerates, more and more energy must be expended on pushing air aside. When a sprinter has less air resistance to contend with, as in wind-aided races or races at altitude, a faster speed can be achieved with the same energy expenditure.

Finishing Dip

A sprinter can gain a slight advantage at the finish by bending the torso forward, across the line. Although the sprinter's center of mass continues moving toward the finish at a nearly constant speed, the sprinter's torso can get to the finish sooner if it is bent forward at the right moment. This is a technique that can and should be coached.

In order to stay balanced when bending forward, the sprinter must keep his center of mass as far back in his body as possible, while leaning his upper body far forward. This requires extending both arms back, to keep the center of mass back in the body. (See Figure 48.) If the center of mass is too far ahead of the feet, the runner will not be able to keep from falling headfirst onto the track.

It is important that the finishing dip, which requires bending the body forward unnaturally, not be confused with the natural and uncontrollable body lean of sprinting mentioned earlier. Bending forward does not increase sprinting speed; it only gets the torso to the finish line slightly.

If, after crossing the line, the runner continues to fall forward because his legs cannot keep up with his upper body, he should swing his arms in an overarm, windmill fashion,

Figure 48. Close races can be won by using a good finishing dip, with trunk bent far forward and arms extended backwards.

creating a secondary axis through the shoulders. (See Figure 49.) This technique, which helps the upper body to rotate backwards and keep the runner from falling forward on the track, should also be coached.

Selecting Starting Blocks

In an effort to simplify their design, many American manufacturers have been producing starting blocks that are too confining, or, in some cases, are disadvantageous for the people who use them.

Figure 49. After bending the body far forward for the finishing dip, balance can be regained by windmilling the arms.

All blocks can be adjusted forwards and backwards, of course, but few have been designed so that they may be moved sideways, an adjustment that is very important to the many sprinters and hurdlers who require a wider base for their fastest possible starts.

Also, many sets of starting blocks in America have been designed with both block faces at the same angle. Studies in Europe have shown that crouch starts are more efficient when the front block is at a 30-degree angle and the back block is at a 60-degree angle. This discovery has led some track and field equipment manufacturers to begin designing triangular blocks, each with one 30-degree, one 60-degree, and one 90-degree angle. Thus, the front block–either right or left–could be rotated into the 30-degree position, and the back block could be rotated into the 60-degree position.

There appears to be slight advantage, too, in providing a way to allow the spikes to be recessed into the faces of the blocks, in order to be able to apply the force against the block face with the sole of the shoe, rather than with the spikes (Such block faces have been manufactured in Europe for years.)

Finally, it should be pointed out that a block which is designed to brace the heel of a foot, so that the heel is not allowed to move backwards at the start, does *not* provide an advantage; it can provide only a disadvantage! When the gun fires, the heel must be able to be depressed in order to put the calf muscles on stretch. This preliminary stretch leads to a much more powerful contraction of the calf muscles a split second later.

Coaching Pointers

• 1. To increase the ability to apply force against the blocks and to increase stride length naturally, do a variety of heavy leg-strengthening exercises, particularly in the off-season.

a. Plyometric exercises (exercises that invoke the stretch reflex in the leg muscles) are highly recommended.

–fast hopping on one leg over distances of 30 to 100m.

–fast hopping on both legs over distances of 30 to 100m.

–long bounding jumps over distances of 30 to 100m.

–double-leg take-offs over ten low hurdles placed 6 feet apart.

–hopping onto and off low boxes.

–hopping upstairs and downstairs.

–depth jumping from the height of a low hurdle and exploding upward.

b. Weight-machine exercises which strengthen the legs,

primarily leg presses with near-maximum poundages, are also effective for greatly increasing leg strength.

c. Barbell exercises that strengthen the legs can also be used. The best of these is step-ups to an 18" bench, with the weight supported on the shoulders. Great poundages can be handled, but it is not recommended that the sprinter ever work to absolute maximum. The weight should be increased only as long as the exercise can be done reasonably quickly and efficiently.

d. Running against resistance—either running or skipping with weights carried on the shoulders, dragging an object while sprinting, or running uphill or up bleacher steps—can be very beneficial in the leg-strengthening program.

• 2. Determine a medium block setting that allows knee angles of approximately 90 and 120 degrees when in the "set" position.

• 3. Practice starts, emphasizing quick forward movement of the leading knee and the opposite arm.

• 4. Do not worry about the degree of lean when coming out of the blocks. Forward lean is the result of acceleration and air resistance, not conscious effort.

• 5. When practicing sprinting, also practice bending the torso forward, as if finishing a close race, followed by windmilling the arms forward to keep from falling forward on the track.

13 ENDURANCE RUNNING

Endurance is the ability to withstand or prolong fatigue in order to complete a particular distance in the shortest possible time. The longer the race, the greater the need for endurance.

The endurance necessary for successful distance running is developed through conditioning programs based upon principles of exercise physiology, an area of science that does not fall within the scope of this book. However, since a runner's energy (the capacity to do work, which is defined as force x distance) is dependent upon aerobic and anaerobic metabolism, a brief overview of the basics of conditioning is included here.

The limiting factor in endurance running is the runner's aerobic capacity–his body's ability to absorb oxygen. During steady running, oxygen is continuously taken into the body's cardiorespiratory system, which delivers it to the muscles. As long as the pace is such that the oxygen supply is adequate, the runner continues to function on the aerobic (oxygen dependent) system. Fats and carbohydrates serve as fuel during aerobic metabolism.

When the racing pace is so fast that the supply of oxygen cannot keep up with the body's aerobic demands, the aerobic system becomes overloaded and the less efficient anaerobic (oxygen independent) system takes over. The anaerobic system is able to use only carbohydrates for fuel, and it produces lactic acid in the muscles as a byproduct.

The shorter races (sprints, hurdles, etc.) are almost entirely anaerobic, while the very long races are almost entirely aerobic. Even the 800m run may be as much as 65% anaerobic, while the 10,000 is less than 10% anaerobic, with the great majority of the anaerobic activity taking place during the closing stages of the race.

Because of the fatigue-producing lactic acid which accumulates in the muscles during anaerobic metabolism, fast running can be maintained for only brief periods of time. Thus, the longer the race, the slower the pace must be.

In order to develop the endurance necessary to improve performance, the distance runner's conditioning program must include a regular program of running that will challenge and overload both the aerobic and anaerobic systems, with much greater emphasis on the aerobic system.

Coaching Pointers

• 1. A training program designed to increase the body's ability to absorb and use oxygen (the capacity of oxygen uptake) should be followed on a year-round basis.

• 2. Additional training to increase the body's ability to function during periods of anaerobic activity should be included in the program.

14
HURDLING

In any hurdling action, the hurdler's objective is to raise the center of mass as little as possible, to spend as little time as possible in the air, and to return to the ground in a position to continue sprinting at the greatest possible speed.

High Hurdling

As was mentioned in Chapters 11 and 12, there are only two ways a runner's speed can be increased—by increasing stride length or by increasing stride frequency. The same is true in high hurdling, except that stride length can only be increased between hurdles with disastrous results. Unless the hurdler is able to reduce the number of steps between hurdles (which of course he cannot do, unless he is taking too many steps to begin with), increasing stride length is not possible. And, since it has been shown that stride frequency is largely an inborn characteristic which cannot be improved appreciably in the mature athlete, one might conclude that it is not easy to bring about improvements in high hurdles times.

However, there are ten unorthodox strides in a high hurdle race, and it is the improvement of stride frequency during those ten strides over the hurdles that brings about an improvement in times. While improved leg strength is the most important factor in the improvement of sprint times, improved clearance technique is the most important factor in the improvement of hurdle times.

The Takeoff. As the hurdler leaves the ground to clear each barrier, the takeoff angle is determined by horizontal velocity, plus the vertical velocity which must be added in order to get the runner over–rather than through–the barrier.

One of the hurdler's goals must be to try to reduce the amount of vertical velocity necessary to clear the hurdle. The less vertical velocity that must be added by the hurdler, the less reduction in horizontal velocity, the lower the takeoff angle and, of course, the less time that must be spent in the air.

Vertical velocity at takeoff can be reduced in two ways–by improving the efficiency of hurdle clearance by lowering the center of mass in the body, or by raising the center of mass at takeoff (which would require an increase in the hurdler's height, which, of course, is not possible). The lower the hurdler's center of mass can be kept during hurdle clearance (or the higher it happens to be at takeoff), the less vertical velocity required.

If the hurdler has the good fortune to be tall, the vertical velocity required for hurdle clearance is automatically reduced. The tall hurdler (whose center of mass is automatically higher at takeoff,) does not have to add as much "speed-reducing" vertical velocity as short hurdlers do.

Trunk lean. The movements and positions of the upper body during hurdle clearance are extremely important in the improvement of hurdling efficiency.

The lower the position of the trunk during hurdle clearance (and thus the lower the center of mass), the less vertical velocity necessary for clearance. The hurdler with less trunk lean must jump higher and spend more time in the air than the hurdler who keeps his trunk forward and low during hurdle clearance.

The trunk-head combination is also important because it

and the leading leg are tied together in action-reaction once the hurdler is in the air. When the trunk and head are lowered, the lead leg rises; when the trunk and head are brought up, the lead leg goes down. (See Figure 50.) The actions of the upper body and the reactions of the leading leg are equal and opposite. Therefore, the hurdler must bring the trunk and head forward and downward during clearance in order to bring the lead leg up over the barrier; then he must snap the upper body upward after clearance to bring the lead leg back to the ground quickly.

Figure 50. The action of raising the upper body takes the lead foot back to the track quickly in reaction.

The trunk also adds greatly to the hurdler's stability during clearance. When the trunk is leaning forward, rotary inertia around the hurdler's vertical axis is increased because more of the hurdler's mass is distributed away from the axis. Since the degree of reaction to a rotary action is always inversely proportional to the rotary inertia involved in the two movements (the greater the rotary inertia of the part involved, the less the reaction), the forward-leaning body reacts less to the various horizontal actions of the arms and legs. The effects of minor errors in arm action, for example, may be reduced significantly if the trunk is leaning far forward during hurdle clearance.

Position of the head. Although there are not many hurdlers who practice it, there is a definite advantage in keeping the head down at the peak of hurdle clearance. When clearing the hurdle, there is really nothing worth looking at, yet most hurdlers feel compelled to look ahead to the next hurdle, which requires keeping the head up.

Lowering the head lowers the hurdler's center of mass slightly, and as has been mentioned earlier, the lower the hurdler's center of mass during hurdle clearance, the less vertical velocity required at takeoff, the less time spent in the air, and the faster the hurdler can return to the ground. Lowering the head is a technique which definitely should be coached!

Arm action. During high hurdle clearance, the action of the trailing leg and the reaction of the leading arm are in parallel planes around two vertical axes–a lower axis through the hip on the trailing leg side and an upper axis through the shoulders.

There is also a vertical primary axis passing through the hurdler's center of mass as he clears the hurdle, but ideally, there is no rotation around this primary axis. (See Figure 51.) However, if the hurdler's arm movement is such that there is less reaction in the upper body than there is action from the trailing leg, then some of the remaining reaction must be absorbed by the body, causing the hurdler to twist in the air and land off balance.

It is true that the hurdler's forward-leaning trunk increases the rotary inertia around his vertical axis and is able to absorb some of the reaction. However, there is enough rotation caused by the trailing leg during hurdle clearance to require that the arm action provide a reaction that eliminates

Figure 51. The hurdler's arm and leg movements should be such that there is no movement around the vertical axis.

rotation in the entire body. Without a proper arm action, the hurdler must either land off balance, or compensate by learning to take off off balance.

The hurdler's trailing leg contains more mass than either of his arms. Therefore, one or a combination of three techniques must be adopted in order to provide arm action that is equal and opposite to the action of the trailing leg:

1. Use both arms in the reaction by having the leading arm forward and the opposite arm back during the takeoff. (See Figure 52.) Then, during hurdle clearance, sweep the leading arm back and the rear arm forward as the trailing leg moves forward. (Never have both arms forward, since this reduces the potential for upper body reaction to the action of the trailing leg.)

2. Increase the radius of the leading arm during hurdle clearance by keeping the elbow straight and swinging the straightened arm back. (See Figure 53.) This practice is condemned by most coaches, but is practiced, out of necessity, by most hurdlers.

Figure 52. It is easier for the hurdler to provide an equal and opposite reaction in the upper body if the lead arm is forward and the opposite arm is back at takeoff.

3. Increase the arc of the leading arm's backward sweep by having the leading arm in front of the body at takeoff, rather than straight ahead. (See Figure 54.) Most top hurdlers use this technique.

Any of these techniques will increase rotary inertia around the upper secondary axis and make it possible for the upper body to react equally to the action of the trailing leg.

Figure 53. Sweeping the lead arm back in a wide arc in reaction to the action of the forward-moving trailing leg helps keep the hurdler in balance.

Low and Intermediate Hurdle Clearance

Because of the relationship of height of center of mass to hurdle height, low and intermediate hurdles can be cleared much more quickly than high hurdles.

The low or intermediate hurdler may not have to raise his center of mass at all to clear the barriers, unless, of course, the hurdler is quite short. Thus, little if any vertical velocity beyond that required in normal sprinting needs to be added at takeoff, making it possible to reduce horizontal velocity very little and to spend less time in the air during

Figure 54. Swinging the lead arm in front of the body before taking off provides the hurdler with a greater arc through which to sweep the arm back in reaction to the forward-moving trailing leg.

hurdle clearance. Low and intermediate hurdlers who are short obviously are at a disadvantage. They will find it necessary to practice high hurdle technique, which will greatly decrease their chances for success.

The only slowing movement during low and intermediate hurdle clearance (provided the hurdler is tall enough so that his center of mass need not be raised) is the requirement of bringing the trail leg out to the side and over the hurdle, which slows that recovery step considerably when it is compared with the sprinting recovery stride.

Because centers of mass do not have to be as low in the body during low and intermediate hurdle clearance as they do in high hurdling (unless the hurdlers are short), a pronounced forward lean of the trunk and a low position of the head are not necessary. (See Figure 55.)

Figure 55. Low and intermediate hurdlers do not have to be as concerned about hurdle clearance technique as do high hurdlers.

Hurdle Selection

In all track & field rulebooks, specifications for hurdles require that they be designed with a certain minimum and maximum pull-over resistance at each height. (Pull-over resistance is defined as the resistance to a steady force applied horizontally to the top edge of the hurdle rail.)

This required resistance is great enough to discourage hurdlers from deliberately knocking hurdles down to run a faster race, yet low enough to keep them from injuring themselves should they accidentally hit hurdles during a race.

In high school, every hurdle, no matter what its height, may have a maximum pull-over resistance of no more than 8 lbs. 13 oz., and a minimum pull-over resistance of 8 lbs. for 30", 7 lbs. for 33", and 6 lbs. for 36" and 39" hurdles. Because of the range of pull-over possibilities for high school hurdles, they can be designed to adjust from the 30" low hurdle height to the 39" high hurdle height without making a corresponding adjustment in the distribution of the weight at the base. When a high school hurdle is adjusted to its low hurdle height, the pull-over resistance is very near the maximum of 8 lbs. 13 oz.; at the high hurdle height, it is near the minimum of 6 lbs.

This is not possible with hurdles designed for collegiate or international competition, all of which must have a pull-over range of 8 lbs. to 8 lbs. 13 oz. Any change in hurdle height—even a 3" change, requires a change in the distribution of base weight in order to keep the pull-over resistance between 8 lbs. and 8 lbs. 13 oz.

Unfortunately, some of the hurdles munufactured in the United States have not been legal, and more unfortunately, some meet directors have not bothered to check them. World records have been set over hurdles that allowed the hurdlers to "run low," because the pull-over resistances were too low.

Any hurdle that does not tip all the way over when a steady force is applied to the top edge of the rail is illegal, and can be dangerous. Any hurdle that can be adjusted to different heights without a corresponding leg weight adjustment for each height is illegal, except in high school.

Before buying hurdles, check a sample of the hurdle with a scale. Press the scale horizontally against the top edge of the hurdle rail and continue pushing until the hurdle tips over. The highest reading on the scale must be between the legal limits or the hurdle is not legal.

Coaching Pointers

• 1. When practicing hurdle clearance, keep the trunk, head and arms as low as possible. Raising any body part increases the time that must be spent in the air.

• 2. When approaching each hurdle, practice positioning the lead arm across the front of the body before leaving the ground.

• 3. While clearing each hurdle, sweep the lead arm out and back, perhaps even into the next lane, in order to maintain balance.

• 4. As soon as the trailing foot has cleared the hurdle rail, quickly raise the trunk, an action which will decrease the time required to get the lead foot back to the ground.

15
RELAY RACING

A relay team is not necessarily the sum of all of its parts. Baton passing skill and placement of runners can produce a relay time that is as much as 3 seconds faster than the total of the best times of the individual runners.

In any relay race, it is the time required for the baton to complete the race that is important—not the times of the individual runners who share in carrying it. Each of the three outgoing runners completes a distance of acceleration in which the baton is not involved, plus there is the "free distance" that can be gained during each exchange.

The importance of free distance, particularly in the 4 x 100m relay, cannot be overemphasized. If the baton exchanges are properly executed, more than one meter (approximately four feet) of the race may be covered by the baton that is not covered by the runners. (See Figure 56.) Although the baton travels 400m in the 4 x 100 relay, the runners have to run less than 397m in the same time—a great advantage in a close race.

The 4 x 100m Relay

The shorter the relay race, the more important the baton exchanges. In the 4 x 100, it is obvious that the baton should be exchanged at the fastest speed the incoming runner is capable of maintaining, and that the runners are as far

Figure 56. Free distance is the distance between relay runners when the baton is exchanged.

apart as possible when the baton is exchanged.

To insure that the baton is traveling as fast as possible at the time of the exchange, the exchange should be made deep in the zone, but not so deep as to risk running out of the zone before the exchange is completed. (In rare cases in which an incoming runner is markedly slower than the outgoing runner, the exchange should be made in the *front* half of the zone.)

Assuming the exchanges in a 4 x 100 are to be made 4½m (almost 15 feet) from the end of the zone, with a minimum of 1m of free distance between the runners (the outgoing runner is 4m from the end of the zone; the incoming runner is 5m from the end), then the runners will cover the following distances:

	Accelerating Without the Baton	Racing With the Baton
Runner #1	0 meters	105 meters
		1 meter of free distance
Runner #2	26 meters (10 + 16)	99 meters
		1 meter of free distance
Runner #3	26 meters (10 + 16)	99 meters
		1 meter of free distance
Runner #4	26 meters (10 + 16)	94 meters
		400 meters

Placement of runners. When selecting relay personnel, it can be beneficial to conduct time trials over three different distances, with each candidate running all three races for time.

Race A–Run 105m (the distance runner #1 carries the baton). Time from the start to the finish.

Race B–Run 125m (the distance run by runner #2 and #3). Time from the 26m mark to the finish (the distance runners #2 and #3 carry the baton).

Race C–Run 120m (the distance run by runner #4). Time from the 26m mark to the finish (the distance runner #4 carries the baton.)

The coach should then find the combination of one A, two B's, and one C that adds up to the lowest total time. Don't be surprised if you find that your fastest runner should be leading off and your slowest runner should be anchoring, since the leadoff runner carries the baton 11m farther than the anchorman.

Another consideration when selecting the relay team should be the particular skills necessary for running each leg. The coach may have to eliminate runners from positions for which they lack a particular skill that is necessary for success in that position:

1. A poor starter should not run the leadoff leg.

2. Poor curve runners should not run the first or third legs.

3. Poor baton "givers" should not run the first, second, or third legs.

4. Poor baton "receivers" should not run the second, third, or anchor legs.

Although many of these are obvious, it is also obvious that if the fastest sprinters would be eliminated from the team because of these rules, the coach will probably bend the rules and place the best sprinters where they have the least chance of hurting the team's chances for success.

Determining sprint go-mark distances. Trying to determine the go-mark distances that will produce the fastest, most efficient baton exchanges can be a frustrating experience for the coach. The most common method is trial-and-error, beginning with an arbitrary go-mark placed on the track some distance ahead of the runup zone. With repeated trials during practice sessions, the mark is adjusted farther away if the incoming runner continues to overtake the outgoing runner, and closer if the incoming runner isn't able to catch him.

While attempting to determine the ideal go-mark distance during practice, the incoming runner usually runs a short distance before passing the baton. Running the entire distance would be so fatiguing that no more than one or two exchanges could be made. Because of the short runup, the practice exchanges are usually made with the incoming runner accelerating, not decelerating as must be the case in an actual race. Also, the fatigue of repeated exchanges during practice, particularly if they are done at the conclusion of the practice session, can have an effect on the placement of the go-mark distance. Obviously, repeated practice over many days, conducted under as near-to race conditions as possible, is necessary to ensure successful baton exchanges in competition.

An alternate system of determining go-mark distances for any two sprint relay runners may be employed by using a simple formula. If the outgoing runner (B) uses the entire 10m runup and the exchange (A to B) is to be made 4½m from the end of the 20m exchange zone with 1m of free distance between the runners, the formula is:

$$G = \frac{75\ (B - A)}{A}$$

G = Go-mark distance in feet

B = Outgoing runner's time for the first 26m of the race

A = Incoming runner's time for the final 25m of the race

If, for example, the incoming runner's time for the final 25m of his 100m leg is 3.0 seconds, and the outgoing runner's time for the first 26m of his leg is 3.6 seconds, then:

$$G = \frac{75\ (3.6 - 3.0)}{3.0}$$

$$G = \frac{75 \times .6}{3.0}$$

$$G = 15 \text{ feet}$$

If the coach wants to assure safer (but slower) baton passes, with the exchanges made near the center of each 20m zone, then the formula changes:

$$G = \frac{60\ (B - A)}{A}$$

G = Go-mark distance in feet

B = Outgoing runner's time for the first 21m of the race

A = Incoming runner's time for the final 20m of the race

When using these formulas, be aware that errors in timing will produce errors in distances. Therefore, it is essential that the timing be as accurate as possible, and that a number of different races be timed. Even if there are slight

Figure 57. The outgoing runner looks back during the 4 x 400 baton exchange, judging the incoming runner's rate of deceleration.

timing errors, the formula will at least establish a point from which to begin the trial-and-error process.

The 4 x 400 Relay

Although the 4 x 400 is still a sprint race, modifications in baton passing techniques are necessary because of increased fatigue on the part of the incoming runners.

First, a visual exchange must be adopted so the outgoing runners can judge the rate of deceleration of the incoming runner and adjust his own acceleration to dovetail with it. (See Figure 57.)

Second, differences in finishing speeds will determine how far into the exchange zone the baton should be passed. A strong finish will allow the exchange to be well into the zone. A weak finish will require a shorter go-mark distance and a much earlier exchange.

Coaching Pointers

• 1. Spend sufficient time determining the best runner placement.

• 2. Use a baton pass that allows as much free distance as possible.

• 3. Practice baton passes regularly, emphasizing the importance of free distance.

PART III

THE JUMPING EVENTS

16
BASIC BIOMECHANICS OF JUMPING

The jumping events can be divided into two general categories—the vertical jumps (high jump and pole vault) and the horizontal jumps (long jump and triple jump).

In the high jump, the primary factors that determine how high the athlete will jump are the velocity and angle of takeoff and bar clearance efficiency.

In the pole vault, they are the velocity and angle of takeoff, the effects of energy storage and the vaulter's movements while holding the pole on vertical velocity, and bar clearance efficiency.

In the horizontal jumps, the primary factors contributing to distance are the velocity and angle of the jumper's takeoff (each takeoff in the triple jump), and the efficiency of the landing position.

Ground Reaction

In all jumping, movement from the ground is the result of ground reaction forces that are equal and opposite to the forces applied against the ground. The greater the forces applied to the ground, the greater the forces returned to the jumper.

The jumper's horizontal velocity at takeoff is the result of a series of horizontal ground reaction forces during the acceleration portion of the runup. The vertical velocity results from the forces applied to the ground during takeoff (in vaulting by the conversion of horizontal to vertical velocity by the pole).

Takeoff Angles

The horizontal jumps require more horizontal than vertical velocity at takeoff and, thus, a lower takeoff angle. The high jump requires more vertical than horizontal velocity (although very little more, since high jump takeoff angles are seldom much greater than 45 degrees), and, of course, the takeoff angle is higher.

The pole vault, although classified as a vertical jump, requires great horizontal velocity and a takeoff angle similar to that of long jumping because of the effects of the pole. The pole, which bends because of the vaulter's weight and horizontal velocity, stores the energy and returns it to the vaulter vertically as the pole unbends.

Flight Curves

Each jumper's center of mass follows a perfect parabolic curve (except for the minimal effects of air resistance) once the jumper is free in the air. The curve for long and triple jumpers is long and low, the result of a greater horizontal than vertical velocity at takeoff. For high jumpers, the curve is shorter, and, of course, much higher. For pole vaulters, the curve, which does not begin until the vaulter has left the pole, is higher yet.

17
HIGH JUMPING

Of the three factors which contribute to successful high jumping, the distance the center of mass can be lifted from takeoff to the peak of the jump (the result of a good takeoff) is by far the most important from a coaching standpoint. The height of the center of mass at takeoff actually contributes the most to the jump—about 2/3 among experienced jumpers—but that factor is entirely dependent upon the jumper's natural physique and the position of his arms and free leg at takeoff.

The Approach

The approach run takes the jumper to the point of takeoff, allows him to assume the takeoff position, and establishes the horizontal velocity for the jumper's flight path after takeoff.

The length of the approach is usually dependent upon the ability of the individual jumper. The beginner, who does not require as much approach speed as a seasoned jumper, should use a shorter runup—6 to 8 strides. The veteran jumper may use as many as 12.

Although some jumpers have attempted straight approaches to the crossbar, it has been shown that a curved approach requires the jumper to lean into the curve, which offsets the otherwise natural tendency to lean toward the bar at takeoff (see Chapter 10). This insures a more vertical takeoff, and produces additional force against the ground.

The Takeoff

When the high jumper leaves the ground and is free in the air, the combination of forward-upward velocity when he leaves the ground and the force of gravity causes his center of mass to follow a parabolic curve. The depth of the curve (the distance from takeoff to landing) is largely determined by the jumper's horizontal velocity at takeoff; the height is determined entirely by his vertical velocity.

It is obvious that the jumper's center of mass must be projected high enough at takeoff to get the center of mass over the bar, and that the peak of the parabolic curve be directly over the bar. No matter how efficient the jumper's bar clearance style might be, it is worthless if the takeoff has not been good. The jumper's primary concern, then, must be to improve vertical velocity at takeoff, which will in turn increase the height of the parabola. And the only way to increase vertical velocity is to increase the vertical ground reaction forces.

Increasing force against the ground. The forces applied to the ground at takeoff–and the resulting ground reaction forces–can be greatly increased with proper use of the arms and the free leg. As the takeoff foot touches the ground, the jumper should assume as low a position as is efficiently possible, with all extremeties in as low a position as can be managed. (See Figure 58.)

Throughout the takeoff, the jumper accelerates the arms and free leg upward, so that all are high when the takeoff foot leaves the ground. (See Figure 59.) Because the arms and free leg are swinging upward while the takeoff foot is pushing downward against the ground, the result is a great increase in the force applied to the ground, and in the force the ground returns to the jumper. The result is increased vertical velocity at takeoff.

Figure 58. At touchdown, the high jumper's arms and free leg are low.

Top-level high jumpers exert a force against the ground that is as much as four times their body weight. However, it is of no use to be able to exert such a great force if the take-off leg is not sufficiently strong to support the resulting load. For this reason, an important part of the high jumper's training program should be heavy leg-strengthening exercises.

Rotational movements at take-off. In order to rotate the body into an efficient back-down position for bar clearance and to develop rotation over the bar, the jumper must initiate certain rotational movements at takeoff. These movements, which are not really visible to the coach or felt by the athlete, are learned by the jumper through trial and error from the time the event is first practiced.

The vertical forces which contribute to the rotation from the ground that is required for bar clearance automatically reduce the ground reaction forces that contribute to

Figure 59. At takeoff, the high jumper's arms and free leg are high.

height. In other words, a jumper must give up some height-getting ability in order to achieve an efficient bar-clearance position. Thus, all high jumping is a compromise. The best technique requires enough rotation from the ground to lay out and rotate over the bar, but not any additional rotation that will reduce the height of the jump more than is absolutely necessary.

The lower the crossbar setting, the more rotation required of the jumper at takeoff, since the layout position must be assumed in a shorter period of time when the bar is lower. As the jumper's ability increases and the crossbar is raised, it is obvious that the amount of rotation required to achieve an efficient bar clearance position is reduced naturally. This works to the jumper's advantage, since a reduction in rotation automatically increases his height-getting ability.

Bar Clearance

The ideal high jump layout position is one in which the largest possible amount of the jumper's body mass is "draped" below the height of the crossbar at the peak of the jump. (See Figure 60.) Some jumping styles allow the jumper's center of mass to pass closer to the bar than others. In the most efficient jumps, it is even possible to have enough body mass draped below the crossbar level to allow the center of mass to pass under the bar while the jumper is going over it. A jumper using an inefficient bar clearance technique may have to project his center of mass some 6'3" in the air in order to clear 6 feet, while a very efficient jumper might have to project his center of mass only up to 5'11" in order to clear the same 6-foot height.

Figure 60. "Draping" the body over the crossbar allows the jumper to clear higher heights.

Once the jumper is in the best possible bar clearance position (with back arched and extremities and head as low as possible), and the hips have cleared, he bends at the hips and brings his legs up. The action of lifting the legs upward brings about an equal and opposite reaction in the upper body, which had been rotating head-down toward the pit. This action, and the simultaneous reaction, serve a necessary dual purpose—the action raises the legs over the crossbar; the reaction keeps the flopper from landing in a dangerous head-down position.

Coaching Pointers

• 1. To improve "spring" in the jumping leg, and to increase the ability to support the increased "load" on the leg at takeoff, do the same leg strengthening exercises recommended for sprinters, particularly in the off-season. (See page 87.) Besides the usual resistance activities, plyometric exercises are recommended (e.g., depth jumping from a height of 30-45", twice a week).

• 2. As the takeoff foot touches down, the arms should be low; as the takeoff foot leaves the ground, the arms should be high.

• 3. The bar-clearance technique should be one that allows as much of the body to be below the top level of the bar as possible.

18
POLE VAULTING

One of the most technically complicated of all sports activities is pole vaulting. In fact, it is probably easier to be "less than perfect" in pole vaulting than in any other track & field event.

A small alteration in any of a number of variables can have a pronounced effect on the success of a vault. The vaulter might be using the wrong pole for his ability, might have the wrong handhold, might be running too slowly, might take off too far out or too far under, might be off in his timing on the pole, or any one or more of dozens of other possibilities. Then there are the condition changes that are beyond the control of the vaulter, such as headwinds, cross-winds and temperature extremes or runways that are "faster" or "slower" than usual.

Through trial and error, however, vaulters learn to adjust and compensate for the many variables that can hinder performance. They learn they can often recover from most of the adversities, at least at the lower heights.

The Approach Run And Pole Plant

The importance of horizontal velocity at the time of the pole plant cannot be overemphasized. The limits in vaulting are directly proportional to the velocity of the vaulter at the time of the plant.

Using the formula $\frac{V^2}{2g} = h$ the accompanying chart has been constructed to show the effects of velocity at the time of the pole plant on bar-clearance potential.

Velocity in Feet Per Second	Conversion to Height (to the nearest inch)	Add 4 Feet for Height of Center of Mass at Takeoff	Add 3 Feet* for Pullup and Pushoff
30	14-1	18-1	21-1
29	13-2	17-2	20-2
28	12-3	16-3	19-3
27	11-5	15-5	18-5
26	10-7	14-7	17-7
25	9-9	13-9	16-9
24	9-0	13-0	16-0
23	8-3	12-3	15-3
22	7-7	11-7	14-7
21	6-11	10-11	13-11
20	6-3	10-3	13-3
19	5-8	9-8	12-8
18	5-1	9-1	12-1
17	4-6	8-6	11-6
16	4-0	8-0	11-0

Most world-class vaulters are traveling at about 28 fps (about 18 mph) at the time of the pole plant. The top speed of a world-class sprinter is about 36 fps (about 25 mph).

An increase of 1 fps in a world-class vaulter's velocity of 28 fps at pole plant, coupled with a 3-foot pullup and push-off (this is not uncommon) will produce a 20-foot vault. Present techniques and equipment need not be improved. The 20-foot vault can come solely because of an increase in

*It has been established through the study of motion pictures that the vaulter's pullup and pushoff can provide approximately three feet to the height he is able to clear.

velocity at takeoff. The pole used, however, will have a very high buckling load rating, since the vaulter's kinetic energy will be so much greater with the increase of velocity.

The Takeoff And Swing-up

The shock from the sudden planting of the pole in the box is absorbed for the most part by the bending pole and to a lesser degree by the flexed right (upper) arm. The arm straightens, as it must, between the time the pole is planted and just after takeoff. (See Figure 61.)

Coaches are not in agreement as to the function of the left (lower) arm at takeoff. Some believe that the two arms form a couple and that the left arm pushes forward, aiding in the bending of the pole. Others consider the left arm merely a stabilizer, with the pole bend at takeoff coming entirely from the forces applied along the pole via the vaulter's right hand.

The vaulters themselves, who should be able to "feel" what they are doing, disagree as enthusiastically as do the coaches and theorists. Some say they consciously push the left arm forward to help bend the pole; others say there is no such effort.

It is the author's opinion that the vaulter is able to determine the direction of bend by applying perpendicular forces to the pole, but that the amount of pole bend is determined almost entirely by the vaulter's kinetic energy at pole plant and takeoff, and by his movements during the swingup. Kinetic energy ($\frac{1}{2}mv^2$), which gives the vaulter an effective weight considerably greater than his actual body weight, bends the pole. At the point of maximum bend, the vaulter's weight is much less than the force required to keep the pole bent, and the pole unbends.

The vaulter adds to the angle of takeoff and takeoff velocity by jumping forward and upward, accentuating this by swinging his free knee forward and upward. The upward-swinging knee adds to the vaulter's force against the runway during the takeoff. (See Figure 62.)

Figure 61. The right (upper) arm straightens during the takeoff.

Figure 62. The forward-upward swinging knee increases the force against the runway during the takeoff.

After the takeoff, the jumper is faced with a number of complex compromises. The vaulter and his pole describe a double pendulum–one is the vaulter and his pole together, rotating upward around an axis at the bottom of the pole; the other is the vaulter rotating upward around an axis through his hands.

Changes in body position can change the rotary inertia

of both pendulums, in opposite ways. If the vaulter bends his knees to decrease his rotary inertia and increase the velocity with which his body is swinging upward on the pole, the rotary inertia of the pole pendulum is increased, slowing its upward velocity. A delay in bending the knees (or a less extreme bend of the knees) will slow down the vaulter's pendulum, but speed up the pole.

Thus, the vaulter must experiment to discover, through trial and error, the timing that will make the best use of the two pendulums and will insure the highest vaults.

The Pullup, Turn and Pushoff

Once the pole has straightened, the vaulter is still experiencing lift, although vertical velocity has decreased greatly because of gravity.

Much has been said about the pull, turn and push at the end of the vault. During the pull, the vaulter is in a position to exert a great deal of force downward through the pole, and also to begin the 180-degree longitudinal rotation required for efficient bar clearance. (See Figure 63.) He pulls with his arms, striving to keep his legs directly above the hands and as close to the pole as possible.

During the push, however, the position of the vaulter's body and the availability of only one arm (which is not in a position to apply force very effectively), makes the push more of a stabilizing movement than an application of downward force. (See Figure 64.)

Bar Clearance

With proper technique, a vaulter has the opportunity to clear heights that are "seemingly above his ability." It is not uncommon for an experienced vaulter to assume a draped position over the crossbar–as a high jumper might–with his center of mass below the bar. (See Figure 65.) In such cases, he has the opportunity to clear a height above the height to which he was able to raise his center of mass from takeoff.

Figure 63. The vaulter pulls forward to begin his needed 180-degree longitudinal rotation.

Vaulting Pole Selection

The primary factors which contribute to the bending of a fiberglass pole are kinetic energy at takeoff and handhold height. The handhold is determined by the vaulter before he begins his run, but the amount of kinetic energy at takeoff depends upon the vaulter's takeoff velocity. The formula is Kinetic energy = $\frac{1}{2}mv^2$ (m = the vaulter's mass; v = his velocity.)

Figure 64. His 180-degree rotation nearly completed, the vaulter uses the pole for stability as he continues to attempt to push downward.

It must be emphasized that the vaulter's velocity just before the pole is planted is not necessarily his velocity during the pole plant. The vaulter's velocity (and kinetic energy) may be reduced greatly if the plant and takeoff are not executed properly. Thus, it is essential to learn to plant the pole with the least possible loss of velocity.

Every fiberglass pole, no matter how it is manufactured, can be classified by its "buckling load." Buckling load is the amount of load–applied inward along the long axis of the

Figure 65. As in high jumping, draping the body over the crossbar allows the vaulter to clear heights that would not have been possible otherwise.

pole—that will cause the pole to bend. The buckling load which a vaulter applies to a vaulting pole is not just the weight of the vaulter, as is often believed, rather a load due to the combined effects of his weight, his velocity, and the technique he uses. Thus, a 160-lb. vaulter, because of his velocity and the technique he uses, may have an effective weight of more than 200 lbs. as the pole is planted, requiring a pole with a buckling load of nearly 200 lbs. The 160-lb. vaulter's effective weight of 200 lbs. at takeoff is sufficient to bend the pole. (See Figure 66.) However, at the point of maximum bend the vaulter's effective weight is less than 200

Figure 66. The pole is bent when the vaulter's effective weight is more than the pole's buckling load.

lbs. and is thus not sufficient to keep the pole bent. Therefore, the pole unbends, giving back the energy that was stored in it during the plant and takeoff. (See Figure 67.)

If the vaulter decides to hold at 14 feet, the pole he selects should have a buckling load of approximately 200 lbs. at 14 feet. That same pole, however, does not have the same buckling load at other handhold heights. At 13 feet it would have a buckling load of about 230 lbs.—much too stiff for our 160-pounder. At 15 feet it would have a buckling load of only 175 lbs., thus if the 160-lb. vaulter whose effective weight is slightly over 200 lbs. at takeoff holds the pole at 13 feet, it will not bend. If he holds at 15 feet and plants the

Figure 67. The pole unbends when the vaulter's effective weight is less than the pole's buckling load.

pole in his usual manner, the pole will overbend and break.

Every fiberglass pole has a great range of buckling load ratings, depending upon the height of the handhold, ranging from very high numbers at the low handholds to relatively low numbers for the highest. It is obvious, then, that poles cannot be classified by body weight of the vaulter with any accuracy without also allowing for his handhold and his takeoff velocity. The many possible handhold heights, coupled with the difficulty of measuring takeoff velocity (which may change from vault to vault), account for the fact that pole selection has become a rather unscientific procedure.

Pole manufacturers have tried to simplify pole selection

by classifying poles according to a combination of bodyweight, handhold height and an average of many possible takeoff velocities. However, this has not proved satisfactory for vaulters whose takeoff velocities are above or below the average. They still have to select their poles through trial and error methods.

Pole Vault Pit Selection

The most common material used in pole vault and high jump pits today is urethane foam, which is produced by chemical action and is characterized by its porous cellular structure. Foam is ideal for landing pits because it is light, durable, and can be manufactured with a degree of firmness ideal for safe and comfortable landings.

It is believed by many coaches (and purchasing agents) that a foam pit is automatically a "safe" pit.* This is not necessarily the case. Foam pits are safe only:

1. If they are large enough to insure that the falling athletes do not miss the target.

2. If they are high enough to allow an adequate distance for deceleration.

3. If they are soft enough to allow an adequate distance of deceleration, but not so soft as to allow "bottoming out."

To ensure safe, comfortable landings, the pit must be designed to decelerate the athlete. (See Figure 68.) The distance of deceleration determines how much shock the body must endure during the landing process. If the pit material is very hard, the athlete will have to decelerate over a short distance, risking discomfort or even injury. (See Figure 69.) If the material is very soft, he may not decelerate enough, passing through the pit and then "bottoming out" against the ground beneath the pit. In each case, deceleration is over too short a distance for comfort and safety.

*When referring to pits as *safe,* "normal" landing positions are presumed.

Figure 68. Landings are comfortable when there is a sufficient distance of deceleration.

Figure 69. Before soft foam pits were developed, vaulters had to use their buckling legs and arms to provide a distance of deceleration.

The shock of landing depends on two factors—the height of the drop and the distance of deceleration.

G (shock or "G" forces)= h (height of the drop)
÷ d (distance of deceleration)

If a vaulter clears 17 feet and falls 14 feet into a three-foot pit, compressing the foam two feet, we have G=14÷2=7 Gs of shock.

It is difficult to locate information on how much shock the human body can stand in different sports situations, but it is known that the maximum limit for exposure to frontal deceleration, as experienced in head-on automobile collisions, is between 40 and 50G for durations of less than 0.1 second.* Pole vaulters learn, through trial and error, which pits are "comfortable" and which are not. If a vaulter experiences discomfort in landing at lower heights, he may change his technique to begin to use his limbs to help break his fall on successive jumps, or he simply may not want to vault any higher.

If the same vaulter had cleared 17 feet and had fallen onto a harder pit, compressing the foam only three inches, he would very likely have been injured as G would be 14÷0.25 which gives 56 G of shock.

(The old adage is true: "It isn't the fall that kills you; it's the sudden stop when you hit the ground!")

When buying foam for your vault and high jump pits, be sure to select material which will provide the safest possible landings. Do not buy according to foam density, since density is merely the foam's mass per cubic foot. Foams of identical density could have widely different grades of firmness.

Always select foam according to its ILD (indention-load-deflection) range, which is a foam compressibility test

*Harris, Cyril M., and Crede, Charles E., *Shock and Vibration Handbook*, New York: McGraw-Hill Book Company, 1961.

developed by the sponge plastics industry. ILD is measured as the number of pounds required to deflect a 4" thick, 15" x 15" piece of foam 25% of its original thickness, using a circular deflection plate with an area of 50 square inches.

Urethane foam is commonly available with ILD ranges of from 5 to 70, since the furniture industry (the largest single consumer of urethane foams) requires foams with many grades of firmness. However, many of these are not suitable for use in pole vault or high jump pits. Try to select foam only in the 15-20 ILD range.

Coaching Pointers

• 1. Since the single greatest contributor to height in pole vaulting is takeoff velocity, the easiest way to increase the vaulter's potential is to have him run faster with the pole. The same leg strengthening exercises recommended for sprinters (page 87) should be used by vaulters, particularly in the off-season. Upper body strength exercises should also be included.

• 2. When the athlete is able to increase takeoff velocity, the increased kinetic energy will require a stronger pole. As the vaulter improves, be prepared to have him change to a stronger pole during the season.

• 3. Because of the complexities in vault technique, it is important to be prepared to recognize the interrelationships of the various individual movements. A technique improvement in one phase of the vault might have a detrimental effect on another phase.

19
LONG JUMPING

Of the factors contributing to successful long jumping, horizontal velocity at takeoff is by far the most important. Although many long jumpers devote hours of practice time to attempting to increase vertical velocity at takeoff, this is invariably done at the expense of horizontal velocity (and sometimes at the expense of the distance of the jump).

Each of the long jump styles (various forms of the hitch-kick and the hang) is a compromise. If the jumper leaves the board at maximum horizontal velocity, forward rotation is developed, which will tend to put him into a head-first position by the time he reaches the pit. Thus, the jumper attempts to introduce compensating backward rotation at takeoff in order to ensure an economical landing position. However, to do this requires that he slow down, and this reduces the distance of the jump.

Of course, the jumper does not notice the problem. From the time he takes his first jumps as a beginner, he learns, through trial and error, that jumping a certain way gives him the most possible distance for his ability. It does not seem to him or to his coach that he is giving up distance in order to get into a good landing position.

The Takeoff

In long jumping, the athlete begins with great runup and

takeoff velocity, but does so at the expense of vertical velocity. The approach is so fast that there is nothing that can be done at the takeoff board to add vertical velocity that compares with the horizontal velocity, so the angle of takeoff must be considerably below the 45 degrees that some coaches insist is the ultimate goal of the jumper. In actuality, the angle is seldom above 25 degrees.

To achieve the long jumping angle, the horizontal component is determined by the approach velocity of the jumper, of course. The vertical component is determined by slightly lowering the center of mass during the final three strides, then straightening the takeoff leg and extending the ankle joint during the takeoff action.

The leading knee and opposite arm are swung upward briefly at takeoff to increase the vertical force applied to the board. (See Figure 70.) The reaction to the action of the upward movement of the limbs is additional force returned by the board.

The movements which contribute to the vertical component in long jumping must not be so great that they retard forward velocity any more than is absolutely necessary. If a jumper decides that he would like to get more height (as a coach might encourage him to do) by "gathering" more, he must slow down at takeoff (probably unnoticeably) in order to do so. Height is not the important factor in long jumping; continuing forward velocity off the board is. (See Figure 71.)

One of the two ways that rotation in the air, which begins on the ground, is initiated is through horizontal eccentric thrust. When any object is traveling forward and one end of it is stopped, the opposite end continues forward at an accelerated rate, and rotation is begun. If the object leaves the ground during that time, the rotary momentum which has been acquired continues until the object returns to the ground.

The foot of the conventional long jumper is stopped at the board for about a 10th of a second while his upper body moves forward 3½-4 feet. Once the jumper is in the air unde-

Figure 70. The long jumper increases the vertical force against the board at takeoff by swinging the leading knee and opposite arm upward.

Figure 71. Some coaches have their long jumpers practice jumping for height by having them try to hit a hanging towel or sponge with their heads. While these exercises may be interesting and perhaps even enjoyable, they do nothing to improve long jumping performance. Jumping for height, whether it is in practice or in competition, reduces the jumper's horizontal velocity and *shortens* the jump.

sirable forward rotation around the transverse (side-to-side) axis is inevitable because of this horizontal eccentric thrust at the board.

To ensure an economical landing position later in the jump, the jumper must introduce a corresponding compensating backward rotation before his foot leaves the board. Unfortunately, any attempt at adding backward rotation at takeoff, deliberate or not, reduces forward velocity appreciably.

In The Air

Once the long jumper is free in the air, the primary concern must be to assume an effective position for landing in the pit. If the takeoff has been efficient, then natural forward rotation from the board continues throughout the jump, making it difficult to get the legs in front of the body's center of mass at landing. Thus, the jumper must attempt to retard or reverse the body's forward rotation while in the air.

Over the years, through trial and error, long jumpers have developed different ways of slowing or temporarily reversing the undesirable forward rotation that is inevitable following a good takeoff. Assuming a hang position, for example, slows forward rotation, since rotary velocity is decreased as rotary inertia around the transverse axis is increased.

The hitchkick style reverses the forward rotation for a moment because of the movements of the legs and arms around two secondary axes—one through the shoulders and one through the hips. The running action of the legs and the windmilling of the arms combine to slow or even reverse the body's forward rotation. (See Figure 72.) But the forward rotation at takeoff can only be retarded temporarily. It returns as soon as the hitchhicking stops and the jumper prepares to land in the pit.

It is important to realize that the more efficient the technique of slowing or reversing forward rotation in the air,

Figure 72. The clockwise rotation of the windmilling arms and the running legs combine to produce a counterclockwise (backward) rotation of the jumper's entire body.

the faster the takeoff (and the longer the jump) may be. The sail technique requires such a slow down at takeoff, for example, that a great deal of distance is automatically lost. The hang technique (which usually contains some of the hitchkick elements) is much more efficient than the sail.

The most efficient of the conventional long jump techniques, by far, is the hitchkick, which reverses forward rotation, making it possible for the jumper to slow down very little at takeoff. However, hitchkicking is one of the most difficult techniques to master because of the time required in the air to complete each hitch. The jumper must be able to jump reasonably well using some other technique before he can even begin to try to learn an efficient hitchkick style.

Importance of the Arms. From takeoff to landing, there are many ways the arms can be used to improve the distance of a jump.

In the hitchkick, and to a lesser extent in the hang technique, the arms play a much more important role in slowing or reversing unwanted forward rotation in the air than most coaches and jumpers seem to realize. It has long been known that the running action of the legs in the hitchkick and the modified running action in the hang style (bent leg forward; straight leg back) help to reverse, temporarily, the forward rotation of the trunk. But, it is only the movement of the lower part of the legs that affects trunk position; the forward and backward movements of the thighs merely check each other.

The arms, however, can be used in their entirety to help reverse forward rotation, although few jumpers use them well. While each arm is rotating clockwise around a transverse axis through the shoulders, the reaction in the trunk is counterclockwise. (See Figure 73.) Since the arms are farthest from the body's center of mass during the upper half of the clockwise arm swing, the counterclockwise reaction in the trunk is most significant at that time. During the lower half of the arm swing, the arms are close to the body's center of mass, and the resulting backward rotation of the trunk is less significant.

The arms should be as straight as possible throughout each revolution—never bent at the elbows; bending the arms at the elbows on the backswing would only increase undesirable clockwise reaction in the trunk.

Figure 73. Windmilling arms are extremely important in aiding the long jumper to assume an efficient landing position.

Just prior to landing, efficient use of the arms can increase the length of the jump.

1. Begin with the arms in front of the body, as though reaching forward for the pit. Then, sweep the arms down and backwards. (See Figure 74.) The reaction to this action is a rotation of the rest of the body in the opposite direction, which raises the legs at a time when they are rotating downward.

Figure 74. A clockwise, backward sweeping of the arms before landing produces a counterclockwise rotation of the entire body.

2. Hold the arms back as far as possible just before landing. (See Figure 75.) This shifts the center of mass backwards in the body, and since the center of mass is following an unalterable flight curve, the effect is a forward shifting of the entire body–including the feet, which are about to land in the sand.

Figure 75. Holding the arms back increases the length of the jump.

The fallacy of holding the legs up. Contrary to popular belief, strength is not required to keep the legs up prior to landing in the long jump. The coach can have his long jumpers hang from chinning bars, holding their legs straight out, developing the strength necessary to hold their legs up when hanging from a bar, but that strength will have no effect on the jumper's ability to keep his legs up just before landing in the pit. In fact, it requires as much strength to hold the legs down as it does to hold them up when the jumper is free in the air. All parts of the body fall with the same acceleration—32 fps^2. The legs fall at the very same rate as the rest of the body!

What, then, causes the legs to appear to drop during the landing phase of the long jump? The answer is forward rotation. The legs remain in the same relative position, but the entire body rotates forward around the body's center of mass so that the legs appear to be dropping prematurely. (See Figure 76.) The real difference is in the attitude of the entire body—not just the legs.

Figure 76. Forward rotation acquired at the board is neutralized temporarily by arm and leg movements in the air, but continues to affect the jumper until his feet touch the sand.

The Landing

The perfect long jump landing is one in which the jumper's feet are as far ahead of his center of mass as possible without causing the jumper to sit back in the pit. In this particular phase of the jump, forward rotation suddenly becomes necessary. In fact, once the jumper's feet touch the sand the more forward rotation the better.

As soon as the jumper lands in the pit, he should swing both arms forward, rotating his body forward over his feet. The chances of falling back in the pit are greatly reduced if the arms are swung forward in this manner.

Coaching Pointers

• 1. At takeoff, concentrate on distance rather than on height.

• 2. While in the air, use arm and leg movements to slow or reverse unwanted forward rotation temporarily.

• 3. Just before landing in the pit, hold the arms as far back as possible, and then swing them forward immediately after landing.

• 4. To improve velocity and takeoff "spring," do heavy leg-strengthening exercises in the off-season. (See page 87.)

20
TRIPLE JUMPING

As in long jumping, horizontal velocity is the most important single factor in contributing to the total triple jump effort. However, the triple jumper does not concentrate so much on distance at the takeoff board, since the hop phase must be modified in order to ensure the greatest possible distance for the total of the three triple jump phases (hop, step and jump).

Studies have shown that the longest triple jumps are most often produced when the hop and the jump phases are of approximately the same length, and the step phase is slightly shorter. It should be pointed out, however, that the hop seems to the athlete to be shorter than it really is, and the step and jump phases seem longer. The athlete must hold back on the hop phase, contribute a near maximum effort in the step, and put everything that is left into the jump.

The Hop

The takeoff for the hop phase requires a lower angle than in long jumping. A takeoff that is too high creates a heavy landing, making it extremely difficult (impossible in some instances) for the jumper to recover and produce a representative step phase. The lower angle of takeoff also reduces the disadvantageous loss of horizontal velocity that is inherent in higher-angled takeoffs.

Although there is not enough time to produce a complete double-arm action during the hop takeoff, as much forward and upward motion of the arms as possible (along with the added upward-forward swing of the leading leg) should be attempted. (See Figure 77.) The ground reaction to this arm and leg action increases the force returned to the jumper from the board, just as in long jumping.

Figure 77. The arms and lead knee are swung upward during takeoff.

The jumper's natural forward rotation in the air, the result of horizontal eccentric thrust at takeoff, is reversed temporarily during the hop phase by employing a one-half hitchkick with the legs and then a downward-backward arm swing of one-half revolution. (See Figure 78.)

It is more important to blend the hop landing into an efficient step takeoff than it is to strive for distance. Having

Figure 78. The arms begin to sweep backwards during the hop phase.

the arms back on landing moves the body's center of mass backward, of course, which moves the landing foot farther forward before landing, with no additional effort. (See Figure 79.) Also, it puts the arms in a position to swing forward and upward through the greatest possible range during the step takeoff.

The Step

Again, in order to create the greatest possible force against the ground, the arms and the free knee are swung forward-upward. (See Figure 80.) This action is much easier during the step phase than during the hop phase because there is more time available to complete the action fully. To get the maximum benefit, the swinging limbs should be as far

Figure 79. The arms are held back at landing.

back at touchdown, and as far forward (and upward) at take-off as possible.

While the jumper is in the air, his natural forward rotation is reversed temporarily by the downward-backward swing of the arms. (See Figure 81.) Because of the nature of the step phase, however, the legs cannot be used, even in a modified or partial hitching action, to help take up forward rotation.

As in all phases of the triple jump, the most efficient step landing is one in which the arms are held as far back as possible, moving the landing foot farther forward. (See Figure 82.)

Figure 80. A vigorous forward-upward arm swing begins the step phase.

The Jump

The arm and leg movements in the jump phase take-off are identical to those in the step phase, except there is usually a higher angle of take-off because of the reduced horizontal velocity and the jumper's striving for more height. (See Figure 83.)

In the air, the triple jumper must take up forward rotation (just as in the previous two phases) to prepare for the landing, in this case with legs well forward of his body. There is not time enough to execute an efficient hitchkick, and many jumpers do not even manage to execute a reasonable hang during this final phase of the triple jump. The jumper temporarily reverses forward rotation by swinging his

Figure 81. The jumper holds the step position in the air.

arms downward, and by bending his legs as much as possible as they are brought forward for the landing. (See Figure 84.)

Coaching Pointers

• 1. Keep altering the distances of the three phases of the triple jump until the jumper is able to produce a hop and jump of approximately the same distance, with a slightly shorter step in between.

• 2. The hop takeoff must be at a low angle, with the emphasis on technique, not distance.

• 3. The arms should be used throughout the triple jump, as far back as possible at the moment of each landing

Figure 82. Again, the arms are back at landing, prepared for a vigorous upward swing.

and as far forward as possible at each takeoff.

• 4. Plyometric leg-strengthening exercises are highly recommended, particularly in the off-season. (See page 87.) Be sure to include triple jump box drills (hopping on and off low boxes) twice a week.

Figure 83. Every bit of remaining energy is put into the takeoff for the jump phase.

Figure 84. The jumper lowers his arms and bends his knees for landing.

PART IV

THE THROWING EVENTS

21
BASIC BIOMECHANICS OF THROWING

The throwing events can be divided into two general categories—the non-aerodynamic events (shot put and hammer) and the aerodynamic events (discus and javelin).

In the non-aerodynamic events, there are only three factors that determine how far the implement will go—the speed of the implement at the moment of release, the angle of release, and the height of release. In the aerodynamic events, however, besides the speed, angle and height of release, there is also the effect which air resistance has on the implement as it travels through the air—an effect that depends largely on the implement's angle of attack.

In each of the four throwing events, the athlete begins by importing horizontal force to the implement—toward the direction of throw in javelin throwing and conventional shot putting; around the body's vertical axis in rotary shot putting and discus and hammer throwing. Just before releasing, a nearly-vertical force is added. It is a great lifting force in the shot and hammer, but less in the discus and javelin, where release angles are not so high. The summed effect of all the horizontal and vertical forces exerted on each implement determines its speed and angle of release.

It is essential that the various body forces contributing to the throw be exerted in the proper order—timed to build on previous forces—in order to provide the greatest possible velocity at release. As the implement increases in velocity before its release, the parts of the body in a position to contribute must be able to move faster than the implement is already moving, if there is to be continued acceleration. This requires the larger, slower muscle groups of the athlete's body to be brought into play first, followed by the smaller, faster groups as the implement approaches maximum velocity prior to release.

Speed of Release

In all throwing events, speed of release is the most important factor. In fact, a small percentage of increase in release speed will always bring about a greater percentage of increase in distance, if all of the other factors remain constant. However, while the athlete must continually attempt to increase the implement's speed at release, he must avoid increasing one velocity component (horizontal or vertical) without also increasing the other. Otherwise, the angle of release is likely to be too high or too low and the distance of the throw may be reduced, even though the release speed has been increased.

Ground Reaction

The forces that can be applied to the implement–both horizontal and vertical–require resistance from firm ground; as the athlete thrusts against the implement, a counter-thrust is received from the ground. Without the resistance from the ground, the powerful actions of the thrower would bring about equal and opposite reactions within his body and the implement would not travel very far.

When a 200-lb. athlete suddenly pushes against the ground with a force of 300 lbs., the ground pushes back with an equal force of 300 lbs. and the person has 100 lbs. of force from the ground to help him lift something away from the ground, such as a shot, hammer, discus, or javelin.

From this, two important points become apparent:

1. The forces that contribute to the acceleration of a throwing implement can be initiated much more effectively while the thrower is in contact with the ground, but not nearly so effectively while he is in the air;

2. The greater the forces applied against the ground, the greater the forces against the implement.

Forces against the ground can be increased with increases in strength. The greater the strength with which the implement is lifted away from the ground, the greater the force against the ground.

Angle Of Release

No matter which of the throwing events is being considered, there is a particular optimum throwing angle for every attempt, no matter what ability the individual thrower happens to possess. However, it is not necessarily the same angle for each thrower in an event, or even the same angle for an individual athlete's different attempts in the same competition.

The optimum angle for the projection of a missile is 45 degrees–if the point of landing is at the same level as the point of release (see page 32). However, since all of the throwing implements are released above ground level, the optimum angle of release must necessarily be less than 45 degrees. How much less depends upon the height of release, the speed of release and, in discus and javelin throwing, on the effects of air resistance.

Flight Curves

Shots and hammers describe nearly parabolic curves as they travel through the air; discuses and javelins describe aerodynamic curves that would be parabolic were it not for the lift and drag effects of air resistance.

Parabolic curves. The moment a shot or hammer is

released and is free in the air, the entire flight path of its center of gravity is determined. The speed and angle of release and the force of gravity after the release cause the implement to follow a parabolic curve.

As mentioned earlier, release angles are determined by the combination of horizontal and vertical velocities. Once the implement is in the air, the horizontal velocity is unaffected by outside forces (except some air resistance), but gravity gradually slows the vertical velocity to zero and then reverses the process, causing the implement to travel progressively faster and faster as it falls. The result is a parabolic curve. The length of the curve (the distance over the ground from release to leading) is determined to a large extent by the horizontal velocity at release; the height of the curve is determined by the vertical velocity.

Aerodynamic curves. Because of discus and javelin design, air resistance causes them to follow flight curves which are not parabolic. As a discus or javelin sails through the air, the air flowing over the implement moves faster than the air flowing underneath, air pressure is diminished above the implement, and a lifting force is created which helps the implement to sail a greater distance than would have been possible without the aerodynamic design.

For the first portion of discus flight, the angle formed between the plane of the implement and the direction of the relative wind (the angle of attack) is a negative angle and there is no lift. This changes as gravity begins to slow the discus. The last half of the discus flight is marked by a positive angle of attack, and the discus experiences a pronounced lifting.

The javelin's angle of attack is positive throughout most of its flight. Even while descending point first, the javelin is inclined at an angle to the relative wind and is continuing to gain distance as it glides toward earth.

22
SHOT PUTTING

The three factors which determine how far a shot may be put are the speed of the shot at release, the angle of release and the height of release. Of these three, speed is by far the most important. A small percentage of increase in release speed will always bring about a greater percentage of increase in distance, if all other factors remain constant.

The putter begins by imparting horizontal force to the shot–toward the direction of throw in conventional shot putting and around the body's vertical axis in rotational shot putting. (See Figure 85.) Then, just before releasing, a nearly vertical lifting force is added. (See Figure 86.) The total effect of the horizontal and vertical forces applied to the shot determines its angle of release and its speed of release.

Strength and Size

There is no doubt as to the importance of strength in shot putting. Indeed, many successful putters spend more time working on strength development than they spend developing technique.

The putter's specific strength exercises should deal primarily with the larger muscles of the legs, trunk, and arm, with an emphasis on power (force x velocity). As mentioned earlier, the greater the forces applied against the ground (the result of increased weight and strength), the greater the

Figure 85. Both the conventional (glide) putter (left) and the rotational putter begin by imparting horizontal forces to the shot.

Figure 86. A nearly vertical lifting force is added before the shot is released.

forces returned to the shot as it is propelled away from the ground.

It is advantageous for the shot putter to be tall. According to tables devised by the late Geoffrey Dyson*, increasing the height of release from 7 feet to 8 feet (requiring a taller putter, of course), automatically increases the distance of the put 9-15" at various release speeds and angles of projection. (A 6-foot putter releases the shot at a height of approximately 7'0".)

An increase in the putter's mass (usually a result of strength training combined with consumption of high protein food) is also an important factor in shot putting. It is a physiological fact that large muscles generate more force than smaller muscles. Thus, the forces muscles are able to generate during the putting action are directly dependent upon their size; the larger they are, the greater the force.

Angle of Release

The optimum angle for good shot putting is between 40 and 42 degrees. The angle can be plotted by bisecting the angle formed by a line drawn from the shot at release to the eventual landing point, and a vertical line drawn upward from the shot at release. Obviously, the greater the velocity of the shot at release, the higher the release angle must be.

Optimum Release Angles When the Height Of Release Is 7 Feet Above Ground

Distance of the Put in Feet	*Optimum Release Angle*	*Distance of the Put in Feet*	*Optimum Release Angle*
25	37°10'	55	41°25'
30	38°25'	60	41°40'
35	39°20'	65	41°55'
40	40°00'	70	42°10'
45	40°35'	75	42°20'
50	41°00'		

*Dyson, Geoffrey H.G., *The Mechanics of Athletics,* London: University of London Press, 1978.

A putter who wants to raise the angle of release on a particular throw must do so by increasing the vertical velocity of the shot before releasing it. (He could also do it by decreasing the horizontal velocity, but the shot obviously would not go as far.) A shot putter who waits until the "arm strike" to try to raise the angle of release finds that making an appreciable change is not possible. The velocity across the circle coupled with the velocity of "lifting" the shot has already effectively determined the angle of release before the arm has had a chance to make a contribution. To raise the angle of release, the putter must increase the lifting action with the legs, trunk and arm during the lifting action.

Conventional Technique

The Glide (Shift!). During the glide across the circle, the emphasis is on horizontal velocity. The putter accelerates the shot as much as possible horizontally (with a little vertical velocity included), and gets into a "lifting" position that will allow the shot's vertical velocity to be increased greatly before the release. (See Figure 87.)

During the shift, the shot must be accelerated over as great a distance as possible. The farther back the shot can begin its trip (even from outside the back of the circle), the greater the distance and time over which the putter will be able to apply force to the shot (force x time = impulse).

Ideally, the movement of the shot from the back of the circle to release is a continuous accelerating movement. For the shift to have maximum effectiveness, the velocity of the shot must not decrease appreciably when the putter reaches the center of the circle.

Although it is difficult to learn, for best results the putter must get into a low position at the end of the shift and, without stopping, be able to accelerate the shot over the greatest possible vertical distance. The lower the body during the shift (the ultimate lifting position allows a right knee angle of approximately 90 degrees), the greater the chances

Figure 87. The "shift" provides most of the horizontal velocity.

for achieving the optimum angle of release.

The Putting Action (Lift!). During the putting action, the emphasis must be on vertical velocity. The right (back) foot lands at the center of the circle fractionally before the left (front) foot, which lands slightly to the left of the toe-board's center. (The putter is still facing the rear of the circle.) Driving the weight forward so that it rotates upward over the braced left leg aids in the lifting action.

The goal during the lift is to apply the greatest possible effective force against the ground. Obviously, the position of the feet is important in the application of this force. If the left foot is too far to the right (e.g., at the center of the toeboard), rotation of the body during the lift is blocked and thus the hip and trunk rotators cannot contribute effectively. If the left foot is too far left ("in the bucket"), application of force against the ground is reduced considerably.

The lift begins with the legs, followed by the trunk, and finally, when the shot is nearing its greatest velocity, by the arm. (See Figure 88.) Since the shot is moving fast at this point, the arm must be very fast in order to contribute additional force.

Figure 88. The "lift" provides most of the vertical velocity.

All of the lifting happens while the body is turning toward the direction of the put. To increase the velocity of turning, the putter should pull the left arm in, as close to the body as possible, until after the shot is released. (See Figure 89.)

Figure 89. Holding the free arm close to the body reduces rotary inertia around the longitudinal axis and increases rotary velocity before the release.

At release, some putters attempt to keep the left foot grounded; others do not. If the shot is released when both feet are off the ground, as is the case with many world-class putters, the reaction that had previously been provided by the ground must now be absorbed by the body. In such cases, the putters with the greatest mass have a decided advantage over their smaller opponents.

Rotational Technique

From 1956 (when rotational shot putting was first discussed at a national coaching clinic in the United States) until the late 1970s (when coaches began to take the technique seriously) it was generally believed that rotational shot putting allowed a far greater application of horizontal force to the shot than was possible in the more conventional styles. It seemed obvious that a discus-style technique would allow the putter to increase the horizontal velocity of the shot significantly because of the greater distance and time available to apply horizontal force to accelerate it.

In reality, however, it has been found that the shot, unlike the discus, does not spin around the outside of the turning shot putter. In rotational shot putting, the putter does all of the turning as the shot remains near the center of the circle, moving slowly toward the direction of throw. (See Figure 90.) The shot stops in its path across the ring, and even moves backwards slightly just before the putter assumes the delivery position.

The great successes of some rotational shot putters in recent years cannot be attributed to an increase in release speed due to the turning action of the rotational technique. The reason for the successes of the rotational shot putters is an increased release speed made possible by an improved delivery position that is produced at the conclusion of the turning action.

Shot Selection

The rules determine the weight and shape of shots; only the size and the surface texture may vary.

There is a slight advantage in selecting a shot that is smaller in size. Because it encounters less air resistance in flight, the smaller shot will travel slightly farther than the larger shot, with the same effort. For example, a 62-foot effort with a 16-lb. shot that is 4 3/8" in diameter will travel 2 3/8" farther than one that is 5 1/8" in diameter.

International rules allow the 16-lb. shot to be as small as 110mm in diameter (approximately 4 3/8"), which is ¾"

Figure 90. The rotational shot putter does most of the turning, while the shot follows a fairly straight path across the circle.

smaller than the maximum allowed. The 4kg women's shot may be as small as 95mm (approximately 3¾").

In high school competition, the 12-lb. shot may be as small as 3 7/8" and the 4kg shot as small as 3¾".

There is also a slight aerodynamic advantage in selecting a shot with a rough surface. As a shot travels through the air, a low pressure pocket forms behind it, creating drag and reducing velocity somewhat. Because of the effect the type of surface has on the air passing nearest the shot, the size of the low pressure pocket behind the shot is reduced when a "rough" shot is used. Thus, drag is reduced and the shot travels farther. The difference is approximately 4 3/8" in a 62-foot effort.

Coaching Pointers

• 1. In all throwing events, it is important that the athletes be as strong (and have as much body mass) as possible. In shot putting, strength training (concentrating particularly on the legs, trunk and throwing arm) is by far the most important contributor to increasing the velocity of the shot at release.

• 2. If the angle of release is too low (which is often the case), develop a technique in which the putter is in a low enough position to be able to "lift" the shot and raise the release angle.

• 3. Attempt to keep the shot moving (and accelerating as much as possible) from the time the shot begins moving until the moment of release.

23
DISCUS THROWING

The factors which determine how far a discus may be thrown are the speed of the discus at release, the angle of release, the height of release, the implement's angle of attack, and the effects of air resistance on the discus as it travels through the air. Speed of release is by far the most important factor. Because the lift on the discus is proportional to the square of its speed, a small increase in speed at release produces a much greater increase in distance, comparatively.

The thrower begins by imparting horizontal force to the discus, around the body's vertical axis. Then, just before releasing, vertical forces are added. (See Figure 91.) The total effect of all the horizontal and vertical forces applied to the discus determines angle of release and speed of release.

As in shot putting, the mass of the thrower (and the length of his levers) has a great effect on the forces that can be applied to the discus prior to release. A larger discus thrower can exert much more effective force over a greater distance than can a smaller thrower. Also, the longer the discus thrower's limbs, the greater the advantage.

Speed of Release

Although the discus ring is only 8'2½" across, the discus thrower is able to have the discus in his hand for a distance of nearly 30 feet during the discus turn. However, the thrower

Figure 91. The discus thrower imparts horizontal force to the discus, and then adds vertical force before the release.

can accelerate the discus only during the two phases when he has both feet on the ground at the same time—about 40% of the total distance traveled by the discus during the turn.

During the first double-support phase, the thrower begins to apply horizontal force to the discus; during the second double-support phase, the greater part of the horizontal force is applied, but the thrower must also add a great vertical force by straightening the body and lifting the discus away from the ground.

During the first double-support phase, there is an advantage in swinging the arms wide, well away from the body's vertical axis. (See Figure 92.) Then, during the first single-support phase, the thrower should also swing the free leg

Figure 92. The arms should be swung wide during the first double-support phase.

wide. Keeping the free limbs away from the body increases rotary inertia, slowing the turn and providing the thrower with more time and better control of the turn. (See Figure 93.)

Figure 93. During the first single-support phase, the free leg should be swung wide.

The second double-support phase provides the thrower with the throwing position. The legs must be in the position to allow the greatest possible force against the ground during delivery, without wasting the horizontal momentum developed during the first double-support phase. (See Figure 94.)

In order to increase the speed of the discus at release,

Figure 94. The second double-support phase provides the thrower with the throwing position.

the thrower continues to apply both horizontal and vertical forces against the ground. In addition, rotary inertia is reduced by bringing the free arm as close to the body as possible. (See Figure 95.) Bringing the arm close to the body reduces rotary inertia and thus increases the body's rotary velocity, since rotary momentum remains essentially the same.

Ground Reaction

As in the other throwing events, the horizontal and vertical forces that can be applied to the discus require resistance from firm ground; as the discus thrower applies force

Figure 95. The free arm is held close to the body as additional horizontal and vertical forces are applied to the ground.

to the ground during the beginning of the turn and during the delivery, he receives equal counter-forces from the ground. Obviously, the greater the thrower's strength and the greater the forces that can be applied against the ground, the greater the forces that are returned.

There is disagreement among the experts as to the importance of having the front foot on the ground at the moment the discus is released. At one time it was thought that the final force applied to the discus would be lost and distance would be decreased unless the thrower were able to apply force against solid ground at release. However, since the throwing arm contributes very little to the length of the throw during the final delivery, and since almost all of the forces against the ground have been applied prior to release, it now appears that trying to hold the foot down during release decreases the speed of the release more than does having both feet off the ground.

Angle of Release

Optimum release angles in discus throwing range from 34 to 40 degrees. Because of the effects of air resistance, the optimum release angle decreases with an increase in the speed of release. This works out well for the thrower, since it is much easier to increase horizontal velocity (velocity of turning) than vertical velocity (amount of "lifting") in the discus turn. Increasing horizontal velocity without an equal increase in vertical velocity automatically lowers the release angle.

The angle of attack (the angle formed between the discus itself and its direction of motion) of the discus at release is 5-10 degrees, with the plane of the discus lower than the direction of motion. When there is a headwind, the angle of attack should be even lower. When there is a tail-wind, it should be raised slightly.

Preliminary Swings

The preliminary swings are used by the discus thrower as a means of relaxation and to get him mentally prepared for the coming effort. The final preliminary swing finishes with the discus well behind the thrower, and with the muscles on stretch. (See Figure 96.) The discus thrower's weight is over his right foot at this point.

The Turn

The discus thrower's rotary momentum around the longitudinal axis–rotary inertia x rotary velocity–must be initiated at the back of the circle, while both feet are on the ground. Since the thrower will not have both feet on the ground again until the delivery begins, this rotary momentum

Figure 96. The muscles are on stretch as the thrower begins the discus turn.

cannot be increased from the time the right foot leaves the ground until the left foot lands in the front of the circle.

Rotary momentum is established by the right foot thrusting eccentrically against the ground as the turn begins. To generate a large rotary momentum and still maintain control, the thrower should initiate the turn with both arms away from the body (increasing rotary inertia) and should swing the right leg wide throughout the turn.

Holding the arms away from the longitudinal axis during the turn also keeps the shoulders, arms, and discus well behind the legs and hips, allowing about 180 degrees through which to apply additional force to the discus during the delivery. Incidentally, it is not necessary to coach the thrower to hold the discus back during the turn; holding the discus away from the body automatically keeps it back.

As the thrower turns, the centripetal and centrifugal forces cause him to lean inward throughout the turn. (See Figure 97.) The more rapid the turn, the greater the lean.

The thrower lands in the center of the circle on the right foot, with hips well ahead of shoulders, and shoulders well ahead of the throwing arm and discus. (See Figure 98.)

Figure 97. During the turn the thrower leans inward because of centripetal and centrifugal forces.

The Delivery

To keep from reducing rotary momentum any more than is absolutely necessary at this point, the thrower must quickly ground the left foot at the front of the circle, slightly to the left of the direction of the throw. As soon as the left foot is back on the ground, final acceleration of the discus can begin.

The thrower's weight shifts from over the right leg to over the left during the delivery. At the same time, the trunk and legs straighten, and the thrower concentrates on pulling the discus around his body's vertical axis in one final explosive, lifting motion. (See Figure 99.)

To further increase the velocity of the discus at release, the thrower should pull the left arm in, as close to the body as possible, during the release. This reduces rotary inertia, which increases the rotary velocity of the upper body. The closer to the body the arm can be held, the greater the upper body's turning velocity at release.

The fingers apply the final forces to the discus. If the forces are applied directly along the plane passing through the discus' center of mass, the discus will sail without a wobble. If, however, the forces are applied above or below that plane, the discus will nutate (rotate around more than one primary axis) and will wobble as it travels through the air.

Flight Of The Discus

Air resistance causes the discus to follow a flight curve that is not parabolic. As the discus sails through the air, the air flowing over it moves faster than the air flowing underneath. Thus, air pressure is diminished above the discus, and a lifting force is created which helps the discus to sail a greater distance than would have been possible without the aerodynamic design.

For the first portion of discus flight, the angle of attack is a negative angle of 5 to 10 degrees, and there is negative lift. This angle changes as gravity begins to slow the discus. The last half of the discus flight is marked by a positive angle of attack, and the discus experiences a pronounced lifting.

In order to increase the stability of the discus during its flight, the thrower should also concentrate on creating as much spin (and as little wobble) as possible in the discus as it rolls off the forefinger.

Figure 98. At the completion of the second single-support phase, the thrower prepares to assume the throwing position, with the discus well behind the shoulders and hips.

Figure 99. During the delivery the trunk and legs straighten, the free arm is held close to the body, and the weight shifts forward.

Discus Selection

The exact size, shape and weight of a discus is determined by the rule books, but the distribution of weight within the discus is not. Because of differences in the weight distribution (which cannot be detected by calipers or scales), some discuses can be thrown farther than others, even though released in exactly the same way.

Every discus has a certain rotary inertia, which is determined by the distribution of mass within the discus. If a great amount of mass is concentrated in the center of the discus, it has low rotary inertia; if most of the mass is distributed around its outer edge, it has high rotary inertia.

When a discus is thrown, the thrower applies both translational and rotational kinetic energy to the discus—translational energy for the distance and rotational energy for the stabilizing spin. Because a hollow discus with the weight distributed to the outside has higher rotary inertia than a "solid" discus, its spin continues for a longer time while in the air, allowing it to stay level, gyroscopically. Since the hollow discus continues to spin in the air and does not "peel off" so soon, it sails farther before landing.

While there are obvious advantages in selecting a hollow discus for competition, there are equal disadvantages in selecting a molded rubber or plastic discus. The rubber or plastic implement may be less expensive, and they may be easier to grip than the metal-rimmed, wooden ones, but they won't travel nearly as far. The lead pellets molded into rubber and plastic discuses tend to be concentrated toward the center, providing nothing but disadvantage for competition. A molded discus has low rotary inertia and cannot spin as well in the air. It peels off very rapidly and doesn't go as far.

Coaching Pointers

• 1. In all throwing events, it is important that the athletes be as strong (and have as much body mass) as possible. In discus throwing, strength training (concentrating particularly on the legs, trunk and throwing arm) is an important contributor to increasing the velocity of the discus at release.

• 2. To further increase release velocity, develop a turning technique which begins wide, with both arms and the free leg away from the turning axis, and ends with the free arm as close to the turning axis as possible.

• 3. At delivery, the thrower should concentrate on applying the force directly along the plane passing through the center of mass of the discus.

24
HAMMER THROWING

Speed of release is by far the most important of the factors that determine how far a hammer may be thrown. A 10% increase in hammer head speed at release can produce an additional 40 feet in distance. The other factors are the angle of release and (much less important) the height of release.

The hammer thrower begins by imparting horizontal velocity to the hammer around a vertical axis. During the succeeding turns, however, the hammer head's plane of motion is progressively steepened, increasing its vertical velocity. Then, just before the release, a nearly vertical force is added. The total effect of all the horizontal and vertical forces applied to the hammer determines its speed and angle of release.

The long radius of turning is especially important, since release velocity is the product of turning velocity and the effective radius of turning. Small increases in the effective radius of turning can add significantly to the distance thrown. At 2.3 revolutions per second, for example, an increase of 1½" in effective radius is worth almost 10 feet of increased distance to the throw. A 3" increase in effective radius increases the distance thrown by about 18 feet.

Centripetal and Centrifugal Forces

As the hammer is accelerated during the turns and it continues to gain in velocity, the thrower must exert more and more centripetal (pulling in) force to counteract the

Figure 100. The hammer thrower begins with preliminary swings.

increasing centrifugal (pulling out) force exerted by the hammer. Since it has been shown that these forces can increase the effective weight of the hammer from 16 lbs. to nearly 700 lbs., it becomes obvious that the thrower with great body mass, great strength, and efficient backward leaning (and "sitting") positions has a decided advantage.

Angle And Speed Of Release

It is important that the hammer be released at an angle as near the optimum as possible (approximately 42-44 degrees). Releasing at too low an angle, which is not uncommon among beginners, automatically reduces the length of the throw.

The thrower begins by imparting horizontal force during the preliminary swings. (See Figure 100.) There is some vertical force, too, since the hammer is lower when it is in the back of the circle than when it is in front, but the main emphasis during the preliminary swings is on horizontal movement.

During the turns (usually three), the high and low points become more pronounced as the thrower continues to add horizontal velocity to the hammer. The greatest periods of acceleration are those times when both feet are on the ground, so the right foot should never be allowed to swing wide. Rather, it should be taken back to the ground in the least possible time and distance. (See Figure 101.) Then, during the delivery, the thrower straightens his body, lifting strongly with his legs and trunk, providing additional vertical velocity and raising the angle slightly more. (See Figure 102.)

Preliminary Swings

There are usually two preliminary swings, or complete revolutions of the hammerhead, before the turns begin. As the hammer is gradually accelerated through these two revolutions, the thrower's center of mass is shifted in the opposite direction from the hammer in order to maintain balance between thrower and hammer. During the second preliminary swing, the body movements are more exaggerated than

Figure 101. During each turn, the right foot is taken back to the ground in the least possible time.

Figure 102. The thrower straightens the body and lifts the hammer during the delivery.

during the first, because of the increase in the hammer's velocity and, thus, an increase in the centripetal and centrifugal forces.

Throughout the preliminary swings and the subsequent turns, the angle of the hammer's upward path increases, with the low point of the swing to the thrower's right as he faces the back of the circle.

The Turns

During the turns, the hammer and thrower rotate

around an axis which passes through their common center of mass and the point of contact with the ground, and which moves from the back to the front of the circle.

As soon as the hammer reaches its low point at the completion of the final preliminary swing, the thrower pivots on the heel of his left and the toe of his right foot. (The right foot is held down as long as possible.) Then, as the hammer nears its high point, the right foot is suddenly lifted and, while the hammer is passing through its high point, is brought quickly around the left leg and back to the ground beside the pivoting left foot. This movement keeps the feet ahead of the hips and the hips ahead of the shoulders throughout the turns. (See Figure 103.)

When the right foot is grounded at the completion of each turn, the thrower is in a position to greatly accelerate the hammer. The feet are well ahead of the hammer, of course, and the hammer is on its downward path, creating an ideal situation for increasing the hammer's velocity.

As rotary velocity increases during the turns, the thrower must assume a less erect, "sitting" position in order to establish the centripetal force that counteracts the increasing centrifugal force, and to increase his stability. He accomplishes this by bending his knees and leaning backwards.

The second and third turns are very much the same as the first, except that the backward-leaning (sitting) position becomes even more pronounced as hammer-head velocity increases. (See Figure 104.)

The Delivery

At the completion of the thrower's final turn, the hammer continues to trail behind his body. After the right foot has been brought around and back to the ground, the hammer passes its low point and the body is then in a position to apply a considerable additional lifting force to the hammer. The left leg is straightened and the trunk and legs continue lifting until the hammer is released.

Hammer Selection

The maximum and minimum dimensions for hammers

Figure 103. The feet stay ahead of the hips and the hips stay ahead of the shoulder during the turns.

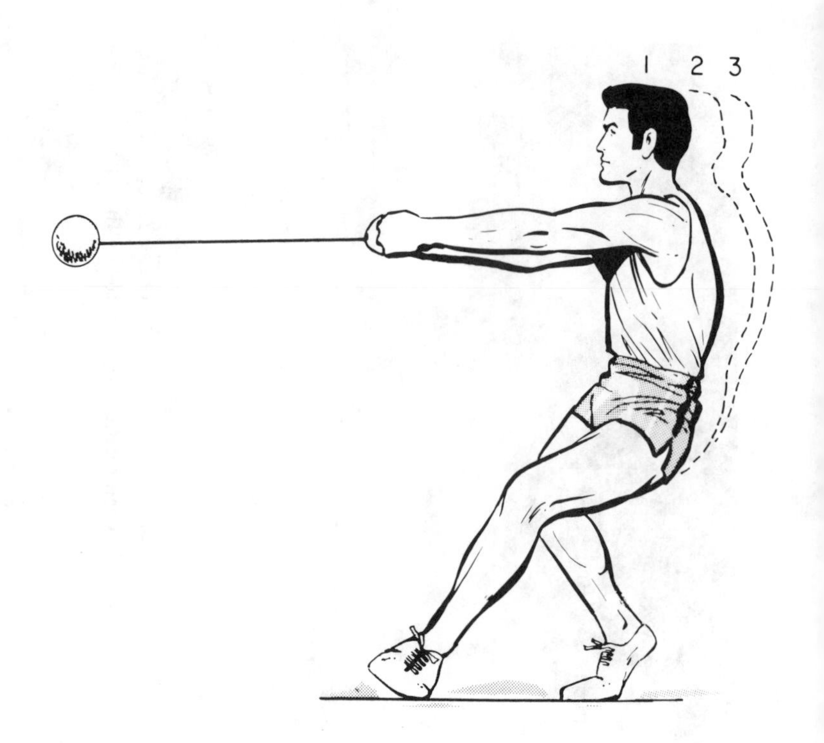

Figure 104. As hammer-head velocity increases, the hammer thrower must lean backward with each succeeding turn.

and hammer heads, as designated in the rule books, allow for two different ways to gain valuable distance. One is in the overall length of the hammer; the other is in the size of the head.

The formula for determining release velocity in hammer throwing is throwing is "Release Velocity = Turning Speed x Effective Radius of Turning." Thus, even if the thrower does not increase his turning speed, his release velocity will be increased if he is able to increase the effective radius of turning. At a turning speed of 2 revolutions per second, an increase of 1½" in effective radius is worth approximately 7 feet of increased distance in the throw. To provide the longest possible effective radius of turning, it is obvious that the thrower should select the hammer with the greatest possible distance between the handle and the head's center of mass.

International and collegiate rules allow the hammer's maximum length (measured from the inside of the grip) to be 3'11¾" (1½" longer than minimum). The size of the head may be as small as 4.33 inches. Also the rules allow the head's center of mass to be ¼" off-center, away from the handle.

By selecting a hammer of maximum length, with the smallest possible head, and the head's center of mass off center by ¼", the thrower's turning radius can be 2" longer than if the minimum length and maximum head size are used.

Coaching Pointers

• 1. In all throwing events, it is important that the athlete be as strong (and have as much body mass) as possible. In hammer throwing, body mass and strength (particularly in the legs, trunk, and arms) are the most important contributors to increasing the velocity of the hammer at release.

• 2. Develop a technique that allows steady acceleration and the longest possible hammer radius during the turns.

25
JAVELIN THROWING

Speed of release is the most important factor contributing to the distance a javelin may be thrown. The aerodynamic lift on the javelin is proportional to the square of its velocity, the same as in discus throwing, making it possible to produce great increases in distance with relatively small increases in release speed. Obviously, then, the javelin thrower's main effort should be to develop a technique that will produce the greatest possible javelin speed at release.

The other contributing factors are the angle of release, the javelin's angle of attack, the effects of air resistance and (to a lesser degree) the height of release.

The javelin thrower begins by imparting horizontal force to the javelin during the approach run, and then contributes vertical and additional horizontal forces during the delivery. The total effect of all the horizontal and vertical forces applied to the javelin determines its angle of release and its speed of release.

The mass of the thrower and the length of his levers have a great effect on the forces that can be applied to the javelin prior to release, as in the other throwing events. However, the thrower must also be able to run fast with the javelin in order to generate the horizontal velocity necessary for long throws. And since the big muscles move more slowly than the smaller ones (and since the implement is not heavy), it is far more important for javelin throwers to be agile than to be bulky.

Speed of Release

The javelin's horizontal acceleration begins when the thrower initiates the approach run. The javelin must be held in such a way that it will not have a slowing effect on the runup, yet will allow the best possible throwing position. (See Figure 105.)

Figure 105. Most javelin throwers prefer to use an above-the-shoulder carry during the runup.

During the release, the velocity of the entire body—not just the throwing arm—is an extremely important and often overlooked contributor to long javelin throws. Velocity of release is measured relative to the ground, not relative to the body. If the javelin is thrown while the thrower's entire body is moving forward, the release velocity is increased. And the faster the body is moving (as long as control is maintained), the greater the velocity of the release.

It is characteristic for beginning throwers to have fast approaches, but then slow down for their throws. Velocity of the javelin throughout the runup is unimportant, except in its contribution to the final release effort. The important thing is the velocity of the javelin at the moment of release.

Of course there is some natural slowdown of the body during the release, even among seasoned throwers, due to the throwing position that must be assumed. But because the thrower's foot is planted and the upper body continues to rotate forward, there is an increase in the forward velocity of the upper body, and of the javelin. (See Figure 106.)

The best throwing position is one in which the thrower has taken a long final stride, increasing backward lean, and, thus, the distance through which the javelin can be pulled. The rear leg is bent slightly and the front leg is straight at the beginning of the delivery, stopping the lower body's forward motion.

An increase in the distance over which the javelin can be pulled during delivery will increase the distance thrown, all other factors being equal. Strength is of the utmost importance to the javelin thrower, of course, but so is range of motion. Exercises that stress pulling against resistance over great ranges of movement, such as in the ideal javelin delivery, should be stressed. Also, javelin technique training should include pulling back the arm as far as possible prior to release.

Angle of Release

There are great differences of opinion among the experts regarding the optimum angle of release. Although it

Figure 106. During the release, the lead foot is planted, causing the upper body and the javelin to increase in velocity as they rotate over the foot.

appears that 34 to 36 degrees is a reasonable average of the opinions, the figures indicate that there have been very successful throws with release angles ranging from less than 30 to more than 40 degrees. Different throwing techniques (and different wind conditions) require slightly different release angles in order to be successful.

The angle of attack (the angle formed between the javelin itself and its direction of motion) at release is usually very close to zero. In other words, the angle of release and the angle of attack are virtually the same. However, the angle of attack may have to be lowered slightly when facing a headwind and raised slightly when there is a tailwind.

Gripping The Javelin

Before beginning the runup, the thrower must get a firm hold behind the back edge of the grip with either the thumb and first finger, the thumb and second finger, or the first and second fingers. The remaining digits should be folded over the top of the javelin, adding stability prior to release and aiding in giving the javelin its stabilizing spin during release.

The Runup

In theory, the faster the approach run, the longer the throw. However, the thrower must sacrifice some velocity to get into an efficient throwing position. As technique improves, the thrower should make every effort to increase his approach velocity.

The withdrawal–drawing the javelin back in preparation for the throw–takes place with 4-6 strides to go. (See Figure 107.) It is timed to coincide with the quick acceleration of the feet, which carries the feet well ahead of the center of mass and puts the thrower's body in a position where it can exert maximum force on the javelin over the longest possible distance. Unfortunately, withdrawing the javelin also tends to slow the approach run. Therefore, the thrower must try to

Figure 107. Withdrawal of the javelin takes place 4-6 strides before the release.

retain as much horizontal velocity as possible during the withdrawal.

As the right foot is grounded at the end of the next-to-last stride, the left foot is already moving ahead of it, with the thrower in a backward-leaning position of some 20 degrees from the vertical. (See Figure 108.) The right knee bends as the left leg continues forward, lowering the body's center of mass in preparation for the throw.

The Delivery

When the heel of the left foot lands, the legs are well apart and the javelin is still held as far back as possible. Some forward velocity is lost at this point; the thrower's goal is to lose as little velocity as possible.

Rotation around two different axes aids in increasing the speed of the javelin during the delivery. The javelin thrower rotates forward over a horizontal axis through the point where his left foot is in contact with the ground and the right shoulder rotates forward around the thrower's vertical axis. Holding the left arm close to the body just before and during the release aids in increasing the velocity of the rotation around the vertical axis.

Figure 108. The thrower assumes a backward lean as the right foot is grounded.

The braced left leg flexes briefly as it lands, but straightens as the javelin is pulled forward powerfully. With chest forward and back arched, the thrower's shoulder begins leading the throw, with the hand far behind. Then, halfway through the delivery, the elbow begins leading, with the hand remaining behind until just before the release. (See Figure 109.)

At release, the fingers give the javelin a stabilizing spin around its long axis, which continues until the javelin has landed. To maintain the javelin's velocity after it has been released, be sure it is thrown with the force applied along the javelin and through its center of mass. This will eliminate wobbling (or the sometimes slightly sideways) deliveries.

Figure 109. The elbow leads the hand until just before the javelin is released.

After The Release

The release of the javelin must be far enough behind the foul line to prevent fouling, of course, yet close enough to insure the longest possible throw for the effort. After the release, the upper body is rotating forward rapidly. The thrower brings the right leg forward quickly, well ahead of the body's center of mass, reversing the forward rotation and braking the body's horizontal motion. (See Figure 110.)

Figure 110. The thrower brings the right foot forward quickly to brake the body's horizontal movement.

Javelin Selection

Javelins are aerodynamic implements designed to sail maximum distances according to the abilities of the particular throwers. As ability increases, the thrower must begin selecting javelins designed to stay in the air longer.

The aerodynamic principles governing javelin flight are very complicated and are not completely understood. However, it is easy to understand one of the aerodynamic factors contributing to the forward rotation of the javelin in the air—the relationship of the javelin's surface area to its center of mass.

Since javelins must rotate into a point-down position before landing, the tail section (everything behind the javelin's center of mass) must have a greater total surface area than the front section. The tail section "catches" the air and the javelin slowly tilts forward. Because it is traveling through the air point first, wind resistance is minimal as the javelin leaves the thrower's hand. But as the javelin begins to slow and more air catches its tail section, the javelin rotates forward around its center of mass.

The greater the distance thrown, the slower the forward rotation in the air must be. Obviously, the javelin that is going to be thrown 250 feet must be designed so that it turns over in the air more slowly than the javelin that will be thrown only 150 feet. Thus, the amount of surface area on the tail of the javelin (the part that catches the air) must be less if the javelin thrower's ability is greater, in order to get the longest possible throw.

This is where javelin design and javelin selection become important. The thrower must select the javelin that will turn over just enough during the time it is in the air so that it lands almost flat, but with the point hitting first. To help solve the problem, javelin manufacturers have designed javelins that are "distance rated," making javelin selection much easier.

Coaching Pointers

• 1. In all throwing events, it is important that the athletes be as strong (and have as much body mass) as possible. In javelin throwing, strength training (concentrating particularly on leg strength and on the full range of motion of the throwing arm) is an important contributor to increasing the speed of the javelin at release.

• 2. To further increase release velocity, develop a technique that includes a fast approach run that continues throughout the delivery of the javelin, and a "pulling distance" that is as long as possible.

• 3. Experiment with different release angles under different conditions to determine the most efficient angle for each javelin and each delivery technique.

BIBLIOGRAPHY

DYSON, GEOFFREY H.G., *The Mechanics of Athletics,* London: University of London Press, 1980.

ECKER, RICHARD E., *Staying Well,* Downers Grove, IL: Intervarsity Press, 1984.

ECKER, TOM, *Track and Field Dynamics,* Los Altos, CA: Tafnews Press, 1974.

ECKER, TOM, *Track and Field: Technique Through Dynamics,* Los Altos, CA: Tafnews Press, 1976.

ECKER, TOM; WILT, FRED AND HAY, JIM, *Olympic Track and Field Techniques,* West Nyack, NY: Parker Publishing Co., 1974.

GAMBETTA, VERN, *Track and Field Coaching Manual,* West Point, NY: Leisure Press, 1981.

HARRIS, CYRIL M., AND CREDE, CHARLES E., *Shock and Vibration Handbook,* New York: McGraw-Hill Book Co., 1961.

HAY, JAMES G., *The Biomechanics of Sports Techniques,* Englewood Cliffs, NJ: Prentice-Hall, Inc., 1978.

HAY, JAMES G., AND REID, J. GAVIN, *The Anatomical and Mechanical Bases for Human Motion,* Englewood Cliffs, NJ: Prentice-Hall, Inc., 1982.

HOPPER, BERNARD J., *The Dynamical Basis of Physical Movement,* Twickenham, England: St. Mary's College Physics Laboratory, 1959.

LICHTEN, WILLIAM, *Ideas from Physics,* Reading, MA: Addison-Wesley Publishing Co., 1973.

McCLOY, C.H., *Mechanical Analysis of Physical Education Activities,* unpublished and undated.

NELSON, BERT, *The Little Green Book,* Los Altos, CA: Tafnews Press, 1983.

SCHROEER, DIETRICH, *Physics and Its Fifth Dimension: Society,* Reading, MA: Addison-Wesley Publishing Co., 1972.

WILT, FRED AND ECKER, TOM, *International Track and Field Coaching Encyclopedia,* West Nyack, NY: Parker Publishing Co., 1970.

WILT, FRED; ECKER, TOM AND HAY, JIM, *Championship Track and Field for Women,* West Nyack, NY: Parker Publishing Co., 1978.

TOM ECKER has coached track and field at the junior high, high school, university, and international levels. The author of 13 sports books and more than 100 articles in national magazines, Ecker has traveled throughout the world, lecturing on various aspects of sport. His humorous and informative talks on biomechanics have taken him before sports groups in 34 states and nine foreign countries.